AN INTRODUCTION TO CHEMICAL ENGINEERING

AN ELEMENTARY TEXTBOOK FOR THE USE OF STUDENTS AND USERS OF CHEMICAL MACHINERY

BY

A. F. ALLEN

B.Sc.Hons., F.C.S., LL.B., late Capt. R.A.F.

WITH 177 ILLUSTRATIONS

1920

PREFACE

THIS book has been prepared for the student of chemistry who has received a fair grounding in that science together with the correlated subjects of physics and mathematics.

Chemistry is one of the most interesting and educative of the sciences, and probably stands alone in that it is practical in the true sense of the word, for many of the operations carried out in the school laboratory are identical in method with, and on the same scale as, similar operations in industry.

Previous to the war, except in the case of our great colleges, chemistry had been almost completely banned from the curriculum, owing mainly to the travesty of the subject that was taught for the purposes of passing stupid examinations set by charlatan educationists. Although the evil they did still lives, there are not wanting signs of a new and healthy growth of interest in this universally operating science.

The author has to thank the several manufacturers for their ready assistance in the preparation of much that is in the book, and F. E. Palmer, Esq., for the execution and revision of various diagrams.

By pointing out errors and suggesting improvements, those who know will be rendering a service to those who are willing to learn.

A. F ALLEN.

1, ELM COURT,
 THE TEMPLE,
  LONDON, E.C. 4.
  April, 1920.

INTRODUCTION

THE student of chemistry realizes very early in his career that he cannot make good progress in his subject without spending a considerable amount of time in the laboratory, doing practical work. A retentive memory may serve to carry a student through a so-called theoretical examination, but without practical experience in laboratory work his value as a chemist is nil, and there is no use for him in the industrial world when his student days have ended.

In the years before the war the outlook for a trained chemist was not a rosy one, and responsible positions were few and far between, but now it may be said that the demand exceeds the supply. The value of the chemist has at last been recognized, and he is being called upon to fill most of the positions which in the less scientific pre-war days were taken by the engineer. The type of chemist demanded is one having a certain amount of engineering training, and it is a sign of the times that the various colleges and institutions throughout the country are taking in hand the question of providing suitable courses in chemical engineering.

It will take some time before courses are standardized, and the student on the threshold of the industrial world will naturally look around for such guidance and information as is available at the moment.

The object of this book is to serve as an introduction to chemical engineering by familiarizing the student with those types of machines which are in general use in the chemical industry. It must not be assumed that the machines put forward represent the acme of perfection from the chemist's point of view; in many cases they are the product of the engineering mind pure and simple, with but scanty knowledge of the laws of chemistry and physics.

The methods of chemical industry may be described roughly as the methods of the laboratory modified so as to permit of operations on a large scale.

The plan of this book has been to take the various pieces of apparatus in common use in a chemical laboratory and to describe their industrial counterparts.

The series of operations involved in the quantitative analysis of a mineral afford a very convenient means of calling to mind the different types of apparatus in daily use by the chemist. These include the mortar and pestle for grinding, the beaker and stirring rod for dissolving and mixing, the filtering apparatus for separating solids and liquids, the evaporating basin for the recovery of soluble solids, the drying oven with its method of temperature regulation for drying solids, the crucible and muffle for high-temperature work, the distillation apparatus for the recovery of valuable solvents, the supply of heat, and the provision of water and steam.

Each of these groups has been treated separately in the following pages, and to them has been added a section on Transport—a point of no account in a laboratory, but one of considerable magnitude in the chemical industry.

CONTENTS

	PAGES
PREFACE · · · · · · ·	v
INTRODUCTION · · · · · ·	vii-viii
LIST OF ILLUSTRATIONS · · · · ·	xiii-xvi

CHAPTER I

CRUSHING AND GRINDING MACHINERY

Jaw Crusher or Nipper—Crushing Rolls—High-speed Crushing Rolls—Sugar-cane Crusher—Rotary Fine Crusher or Cracker—Edge Runner Mill or Chaser—Iron Edge Runner Mill—Granite Edge Runner Mill—Overhead Driven Mill with Revolving Pan — Disintegrators — Use of Screens — Buhrstone Mills—Vertical Runner Mill—Roller Mills—Ball Mills—Pebble Mills—Tube Mills—Combination Tube Mill—Stamps -   -   1–39

CHAPTER II

SEPARATING AND MIXING MACHINERY

The Grizzly—Sieves—The Trommel—Telescopic Screen—Sifting Reels — Centrifugal Dressing Machines — Powder Dresser—Vibration Machines — Newaygo Screen — Shaking Sifter—Gravity Leg Separator—Air Separators—Electro-magnetic Machines — Magnetic Pulley — Water Separation — Settling Tanks — Cane Juice Subsider — Levigating Mill — Mixing Machines—Putty Mill—Pug Mill—Horizontal Mixer—Cone Mill—Batch Mixer—Crutching Machines   -   -   40–69

CHAPTER III

FILTERING APPARATUS

Bag Filter—The Filter Press—Frame Press—Chamber Press—Filter Plates—Filter Cloths—Methods of Closing—Methods of Feeding—Centrifugal Machines—Weston Centrifugals—Types of Lining—Friction Pulley—Water-driven Centrifugals—Interlocking Gear   -   -   -   -   -   70–97

CHAPTER IV

DRYERS AND EVAPORATORS

PAGES

Flue Heater—Steam Heater—Firman Dryer—Rotary Dryer—
Warm-Air Drying—Sturtevant System—Timber Drying—
Triple Drying System—Vacuum Drying—Shelf Dryer—
Vacuum Rotary Dryer—Vacuum Drum Dryer—Passburg
System—Johnstone Dryer—Combined Vacuum Dryer, Mixer
and Ball Mill—Continuous Dryer—Evaporators—Spontaneous
Evaporation — Kettles — Open Pans — The Grainer — Steam-
jacketed Pans—Steam-jacketed Kettles—Tilting Kettles—
Steam Evaporating Pans—Wetzel Evaporating Pan—Vacuum
Pans—Jet Condenser—Surface Condenser—Wet and Dry
Vacuum Pumps—Calandria Vacuum Pan—Multiple-effect
Vacuum Pans—Kestner System—Climbing and Falling Film
Evaporators—Salting Type Evaporators—Multiplex Evapor-
ators   -   -   -   -   -   -   -   98–144

CHAPTER V

DISTILLING APPARATUS

Column Still—Rectifying Still—Continuous Still—Coffey Still—
Extraction Plant—Mineral Oil Plant—Lubricating Oil Plant—
Tar Stills—Retorts—Nitric Acid Retorts—Pot Stills—By-pro-
duct Coke Ovens—Gas Retorts—Dowson Process—Pressure
Plant—Suction Plant—Bituminous Plant—Hydrogen Plant—
Kilns — Chamber Kilns — Rotary Calciner — Cement Kilns—
Muffle Furnace—Reverberatory Furnace—Regenerative Fur-
nace—Roasting Furnaces—Air and Water Cooled Shafts 145–178

CHAPTER VI

WATER TREATMENT PLANT

Hardness of Water — Common Impurities — Scale — Corrosion—
Frothing—Water Softening—Lime-Soda Process—Intermittent
Plant—Continuous Plant—Automatic Apparatus—Permutit
Processes   -   -   -   -   -   -   179–193

CHAPTER VII

THE CONTROL OF TEMPERATURE

Steam Control—Baldwin System—Isothermal Valve—Applications
to Still—Jacketed Pan—Dye Vessel—Vulcanizing Press—Gas
Producers — Gas Heating — Reducing Valve — Refrigerating
Machinery — Lightfoot System — Ammonia System — Carbon
Dioxide System—Ice-making—Can Ice—Cell Ice—Plate Ice—
Cold Storage—Brine Pipe System—Direct Expansion Pipe
System—Air Circulation System—Absorption System - 194–210

CHAPTER VIII

TRANSPORT

PAGES

Conveying Solids—Wheelbarrow—Tipping Waggons—Runways—
Aerial Wire Ropeways—Single Wire System—Double Wire
System—Bucket Elevators—Conveyors—Worm Conveyor—
Scraper Conveyor—Mechanical Raker—Belt Conveyors—
Throw-off Carriage—Apron Conveyor—Bucket Conveyor—
Shaking Conveyor—Grasshopper Conveyor—Conveying Liquids
—Pipes—Anti-corrosion Material—Tantiron—Ironac—Vitreon
—Vitreosil—Ceratherm—Vitreosate—Elevating Liquids—The
Acid Egg—The Air Lift or Pohlé System—Plunger Pumps—
Centrifugal Pumps—Ceratherm Pumps—Conveying Gases—
Pipes — Chimneys — Fans — Radial Flow Fans — Mixed Flow
Fans — Rateau Fan — Rotary Blower — Compressors — Hori-
zontal and Vertical Types—Multi-stage Compressors—Vacuum
Pumps—Roughing Pumps—Siemens Oil Pump—Mercury
Diffusion Pump   -   -   -   -   - 211–255

CHAPTER IX

APPENDIX

Distillation of Liquid Mixtures—Air Compression—Belt Conveyors
—Belting—Shafting—Refrigerating Machines—Low Boiling-
Point Liquids—Freezing-Point of Brines—Freezing-Point of
Calcium Chloride Solutions—Freezing Mixtures—Comparison
of Thermometer Scales—Percentage of Lime in Milk of Lime—
Specific Gravities of Soda Solutions—Useful Data of Common
Substances—Temperature, Pressure, and Total Heat of
Steam   -   -   -   -   -   -   - 256–267

INDEX   -   -   -   -   -   -   - 268–272

LIST OF ILLUSTRATIONS

FIG.   PAGE

1. " STAG " ORE-CRUSHER — 3
2. " STAG " ORE-CRUSHER: SECTION — 4
3. G.A. HIGH-SPEED CRUSHING ROLLS — 7
4. FINE CRUSHING ROLLS, TYPE 5, CLASS 2 — 8
5. FINE CRUSHING ROLLS, TYPE 5, CLASS 1 — 9
6. SUGAR-CANE MILL — 10
7. ROTARY CRUSHER: SECTION — 11
8. IRON EDGE RUNNER MILL — 12
9. GRANITE EDGE RUNNER MILL — 13
10. MILLS WITH REVOLVING PAN — 15
11. DISINTEGRATOR — 17
12. DISINTEGRATOR: SECTION — 19
13. DISINTEGRATOR: SECTION — 19
14. G.A. DISINTEGRATOR PLANT: ELEVATION — 21
15. G.A. DISINTEGRATOR PLANT: PLAN — 23
16. MILLSTONE MILL — 25
17. VERTICAL RUNNER MILL — 26
18. ENCLOSED END RUNNER MILL — 27
19. TRIPLE GRANITE ROLLER MILL — 28
20. BALL GRINDING MILL — 30
21. " ATLAS " PEBBLE GRINDING MILL — 32
22. " STAG " BALL MILL: SECTION — 33
23. " STAG " TUBE MILL: SECTION — 36
24. STAMPS FOR CRUSHING — 38
25. PORTABLE SCREEN — 41
26. TELESCOPIC REVOLVING SCREEN — 42
27. REELS — 43
28. SIFTING REELS — 43
29. POWDER DRESSER — 44
30. CENTRIFUGAL DRESSING MACHINE — 45
31. SHAKING SIFTER — 47
32. SHAKING SIFTER AND CONVEYOR — 48
33. GRAVITY OR LEG SEPARATOR — 49
34. " STAG " AIR SEPARATOR — 51
35. ARRANGEMENT OF ELECTRO-MAGNETS IN SPOUT — 53
36. ELECTRO-MAGNETIC SEPARATOR — 54
37. MAGNETIC PULLEY — 55
38. LEVIGATING MILL — 56
39. CANE-JUICE SUBSIDER — 58

FIG.		PAGE
40. DESIGN OF LEVIGATING PLANT: PLAN		59
41. DESIGN OF LEVIGATING PLANT: ELEVATION		60
42. FLOW SHEET OF MILLING PROCESS		61
43. POSITIVE DRIVEN PUTTY MILL		62
44. VERTICAL PUG MILL		63
45. " POWERFUL " HORIZONTAL MIXER		64
46. CONE PAINT MILL: SECTION		65
47. DOUBLE MIXER FOR SEMI-LIQUIDS		66
48. " OPEN-DRUM " MIXER		67
49. UNDER-GEARED MIXER		68
50. CRUTCHING MACHINE: SECTION		69
51. BAG FILTER		71
52. FILTER PRESS PLATES AND FRAMES		74
53. FILTER PRESS: SECTIONS		76
54. FILTER PRESS: PLATE AND FRAME TYPE		77
55. FILTER PRESS PLATE: RECESSED TYPE		78
56. FILTER CLOTH CLIPS: BAYONET TYPE		78
57. FILTER CLOTH CLIPS: SCREW TYPE		79
58. FILTER CLOTH: FIXING IN RECESSED TYPE		79
59. FILTER PRESS: RECESSED TYPE		80
60. FILTER PRESS: CENTRAL SCREW CLOSING		80
61. FILTER PRESS: RACK AND PINION CLOSING		81
62. FILTER PRESS: COMPRESSED AIR CLOSING		82
63. FILTER PRESS: HYDRAULIC RAM CLOSING		83
64. " WESTON " CENTRIFUGAL BASKET: SECTION		85
65. TYPES OF LININGS FOR BASKETS		86
66. BEARING FOR CENTRIFUGAL SPINDLE		87
67. BEARING FOR CENTRIFUGAL SPINDLE		88
68. CENTRIFUGAL FRICTION PULLEY		89
69. " WESTON " CENTRIFUGAL MACHINE		91
70. " WESTON " CENTRIFUGAL MACHINE: WATER-DRIVEN		93
71. INTERLOCKING GEAR FOR WATER-DRIVEN CENTRIFUGAL: SECTION		96
72. " FIRMAN " DRYER: LONGITUDINAL SECTION		99
73. " FIRMAN " DRYER: CROSS SECTION		100
74. " HERSEY " ROTARY DRYER: CROSS SECTION		101
75. COMBINATION ROTARY DRYER: CROSS SECTION		101
76. G.A. DRYING PLANT: PARALLEL DRIVE		102
77. G.A. DRYING PLANT: RIGHT-ANGLE DRIVE		102
78. RECORDING HYGROMETER		104
79. TYPICAL GUIDE CHART		104
80. HYGROMETRIC CHART		106
81. STURTEVANT TRIPLE DUCT DRYER: SECTION		108
82. VACUUM SHELF DRYER		110
83. VACUUM DRUM DRYER		113
84. VACUUM " JOHNSTONE " DRYER: SECTION		114
85. VACUUM DRYER: MIXER AND BALL MILL		116
86. CONTINUOUS CONE VACUUM DRYER		117
87. EVAPORATING PAN: OPEN TYPE		120

FIG.   PAGE
88. CRYSTALLIZING PAN: STEAM HEATED   -   -   - 122
89. TILTING KETTLE   -   -   -   -   -   - 123
90. ASPINALL STEAM EVAPORATING PAN   -   -   - 124
91. WETZEL EVAPORATING PAN   -   -   -   - 126
92. COPPER VACUUM PAN   -   -   -   -   - 128
93. CAST IRON CALANDRIA VACUUM PAN   -   -   - 129
94. G.A. VACUUM PAN: JET CONDENSER   -   -   - 130
95. G.A. VACUUM PAN: TORRICELLIAN CONDENSER   -   - 131
96. INJECTION CONDENSER: SECTION   -   -   - 132
97. SURFACE CONDENSER: SECTION   -   -   - 132
98. KESTNER CLIMBING FILM SINGLE-EFFECT EVAPORATOR   - 136
99. KESTNER FALLING FILM SINGLE-EFFECT EVAPORATOR   - 136
100. KESTNER QUADRUPLE-EFFECT, ETC.: DIAGRAM   - 138
101. KESTNER SALTING TYPE EVAPORATOR: SECTION   - 139
102. " MULTIPLEX " FILM TRIPLE-EFFECT EVAPORATOR   - 142
103. " MULTIPLEX " TRIPLE-EFFECT EVAPORATOR: SECTION   - 143
104. DIAGRAM OF STILL COLUMN -   -   -   - 146
105. RECTIFYING STILL   -   -   -   -   - 148
106. CONTINUOUS DISTILLATION APPARATUS: DIAGRAM   - 149
107. CONTINUOUS STILL: DIAGRAM   -   -   - 150
108. DIAGRAM OF COFFEY STILL -   -   -   - 151
109. EXTRACTION APPARATUS   -   -   -   - 153
110. SCOTT OIL EXTRACTION APPARATUS: DIAGRAM   - 155
111. LUBRICATING OIL DISTILLING PLANT: ELEVATION   - 156
112. LUBRICATING OIL DISTILLING PLANT: PLAN   - 157
113. " DOWSON " STEAM JET PRESSURE GAS PLANT   - 161
114. " DOWSON " SUCTION GAS PLANT   -   -   - 162
115. 30 H.-P. SUCTION GAS PLANT   -   -   - 164
116. DOWSON BITUMINOUS PLANT -   -   -   - 165
117. ROTARY CALCINER   -   -   -   -   - 168
118. SIEMENS REGENERATIVE FURNACE: DIAGRAM   - 170
119. HARRIS MECHANICAL ROASTING FURNACE   -   - 172
120. " A " TYPE SHAFT FOR ROASTING FURNACE   - 173
121. " B " TYPE SHAFT FOR ROASTING FURNACE   - 175
122. H.H. TYPE MECHANICAL ROASTING FURNACE -   -   - 177
123. RECTANGULAR WATER SOFTENING APPARATUS   - 186
124. AUTOMATIC WEIGHING AND MEASURING APPARATUS FOR
          WATER-SOFTENING APPARATUS -   -   -   - 187
125. POSITIVE DISCHARGE VALVE FOR WATER-SOFTENING APPA-
          RATUS   -   -   -   -   -   - 189
126. CYLINDRICAL WATER-SOFTENING APPARATUS   -   - 191
127. PERMUTIT WATER-SOFTENING APPARATUS: DIAGRAM   - 192
128. " ISOTHERMAL " STEAM VALVE   -   -   - 195
129. " ISOTHERMAL " THERMOMETER   -   -   - 196
130. G.A. " ISOTHERMAL " TEMPERATURE CONTROL APPARATUS - 197
131. " ISOTHERMAL " CONTROL OF STILL -   -   - 198
132. " ISOTHERMAL " CONTROL OF STEAM-JACKETED PAN   - 199
133. " ISOTHERMAL " CONTROL OF DYE VESSEL -   -   - 200
134. " ISOTHERMAL " CONTROL OF EXHAUST STEAM   -   - 200

FIG.		PAGE
135.	"ISOTHERMAL" CONTROL OF VULCANIZING PAN -	201
136.	"ISOTHERMAL" CONTROL OF BLAST FOR GAS PRODUCER -	201
137.	"ISOTHERMAL" CONTROL OF COTTON-SPINNING ROOMS -	202
138.	"ISOTHERMAL" SUPERHEATED STEAM VALVE -	202
139.	"ISOTHERMAL" GAS VALVE -	203
140.	"ISOTHERMAL" GAS VALVE -	203
141.	DIAGRAM OF LIGHTFOOT REFRIGERATION SYSTEM -	204
142.	OPEN CONDENSER -	205
143.	HORIZONTAL AMMONIA COMPRESSOR -	207
144.	VERTICAL CARBON DIOXIDE COMPRESSOR -	207
145.	SIDE-TIPPING WAGGON -	212
146.	END-TIPPING WAGGON -	212
147.	RUNWAY FOR MINE -	213
148.	INTERWORKS TRAFFIC: PORTABLE ROPEWAY -	214
149.	STANDARD FOR SINGLE-ROPE SYSTEM -	215
150.	STANDARD FOR DOUBLE-ROPE SYSTEM -	216
151.	ELEVATOR CHAIN -	218
152.	GRAVITY BUCKET CHAIN -	219
153.	DUST PROOF ELEVATOR -	220
154.	BOOT FOR ELEVATOR -	221
155.	HOOD FOR ELEVATOR -	222
156.	SPIRAL FOR WORM CONVEYOR -	223
157.	SPIRAL CONVEYOR -	224
158.	SCRAPER CONVEYOR -	225
159.	THREE-PULLEY BELT CARRIER -	227
160.	THROW-OFF CARRIAGE -	228
161.	ELEVATOR AND CONVEYOR -	229
162.	SLAT CONVEYOR -	230
163.	GRASSHOPPER CONVEYOR -	232
164.	TANTIRON ACID EGG -	237
165.	TANTIRON HORIZONTAL PUMP -	239
166.	TANTIRON VERTICAL PUMP -	240
167.	CERATHERM BODY IN IRON CASTING -	241
168.	CERATHERM IMPELLER -	242
169.	CERATHERM PUMP: INTERIOR -	242
170.	CERATHERM PUMP: SMALL SIZE -	243
171.	CERATHERM PUMP: SUCTION SIDE INTERIOR -	244
172.	CERATHERM PUMP: PRESSURE SIDE INTERIOR -	244
173.	VITREOSATE THREE-WAY TAP -	247
174.	ROBEY COMPRESSOR -	251
175.	VERTICAL OPEN-TYPE AIR COMPRESSOR -	252
176.	BELT-DRIVEN THREE-STAGE COMPRESSOR -	253
177.	FOUR-STAGE OXYGEN COMPRESSOR -	254

INTRODUCTION TO CHEMICAL ENGINEERING

CHAPTER I

CRUSHING AND GRINDING MACHINERY

GRINDING is one of the most important operations in the industrial preparation of chemical products, because it facilitates the handling of the raw material and shortens the time required for any subsequent reaction, and for the selfsame reason many chemical products have to be put on the market in the form of a powder or paste.

In the selection of any particular machine the nature of the initial material and the result desired are the determining factors, but it is useful to remember that reduction by easy stages is, in the long-run, the most economical of time and money. This fact has long been recognized by makers of grinding machinery, with the result that there are on the market many types of machines designed to deal with all kinds and conditions of material, from the crushing of the hardest rocks to the production of the finest powders.

Jaw-Crusher.—This machine is often known as a "nipper," and is one of the simplest and cheapest crushing machines available. Many forms of jaw-crushers exist, but generally it may be said they are so designed that one part of the machine is stationary, with a corrugated face of chilled iron against which works a similar but movable face or jaw, with a **V**-shaped opening between. The movable plate is moved alternately forward

1

and back by an eccentric on the shaft. These nippers, or jaw-crushers, are made in different sizes, the rock opening varying from 12 by 6 inches to 36 by 48 inches; they weigh anything up to 30 tons, and require from 1 to 100 horse-power. These machines require a very solid foundation on account of the very great strain during working. As a rule a method of adjusting the working parts, such as a screw or similar device, is provided, so that the machine can be set for the production of a rough or fine product.

The jaw-crusher will crush the hardest materials, and is very largely employed for ore-crushing and in the gypsum industry; the rock is broken to somewhat less than the size of a man's fist, or capable of passing a 2¼-inch ring. The capacity of a jaw-crusher, in tons per hour, and the power required vary with the condition of the rock, and as a rule dry rock, especially gypsum, is more easily crushed than wet.

Fig. 1 shows a popular size of the " Stag " ore-crusher, manufactured by Edgar Allen and Co., Ltd., Sheffield, to which the following details refer: Massive cast-iron body with easily renewable bearings. The Pitman is of cast iron of substantial design, with renewable cast-iron adjustable bush for eccentric shaft, and is also fitted with renewable cast-steel toggles to receive the nose toggle plates. The swing jaw is of cast iron, accurately bored and fitted to shaft, and fitted with renewable cast-steel toggle seatings. The jaw faces are of manganese steel, cored out at the back to reduce weight, and spelter is run in to ensure a soft cushion for bedding on the swing jaw. The jaw faces are readily accessible, and can be reversed or renewed by an unskilled labourer. The side plates are of hard cast iron, made in one piece, and assist to hold the fixed jaw face securely in position. The toggle plates are of hard cast iron, held in position by the tension rod, which is fitted with india-rubber buffers. The shafts are made from best hammered mild steel forgings, accurately turned to gauge and polished all over. The flywheels are of cast iron, bored to gauge, fitted and keyed to the eccentric shaft, and turned to receive a flanged pulley. The fast pulleys are of cast iron, flanged

FIG. 1.—" STAG " ORE-CRUSHER.

and bolted to either flywheel, turned and crowned on the face; the loose pulleys of wrought iron have a suitable bush of brass, and must revolve at a rate of not less than 250 to 280 revolutions per minute. Each machine can

be regulated while running to break the material to any
special size, by means of cast-iron wedge blocks, the

FIG. 2.—" STAG " ORE-CRUSHER: SECTION.

output of all machines being calculated for crushing
limestone to pass a 2¼-inch ring.

Fig. 2 is a sectional illustration of the breakers with cast-iron bodies, sizes 20 by 8 inches and 20 by 10 inches.

Crushing Rolls.—This type of machine is of heavy construction, and is designed for reducing the produce of the jaw-crusher to sand. The essential part of these machines consists of two cylinders capable of being given an inward turning motion whereby material as it is fed to them is carried along and crushed between them. As a rule the bearings of one roller are rigidly fixed, whereas the bearings of the other roller are held in position by powerful springs, which not only allow the rollers to be set at a definite distance apart—usually ¼ inch or the size of the finished product—but also allow the rollers to give when any material is fed to the machine which it cannot crush without causing damage to the machine itself. Each roller is generally made up of an outer shell of cast steel, chilled cast iron, manganese steel, or a steel forging, which is fixed to a centre of cast iron in such a way that it can be easily removed and replaced when worn out. Owing to the hard wear on the rollers, they require to be constantly trued-up, an operation which can be effected in the ordinary way or by the use of an emery wheel on the rollers when in position. To diminish the amount of dust or smalls formed, the surfaces of the rollers are often fluted, the pitch of the serrations being varied according to the product required. In some cases the rollers are provided with teeth, when designed for cubing granite, limestone, macadam, etc., and will take pieces up to 6 inches cube and reduce them to 2¼ inches cube.

To obtain a true crushing action the rollers should be driven at the same peripheral speed, but in practice it is found that a small difference in speed reduces the amount of wear very considerably without causing excessive grinding. The speed of the rolls varies from 100 to 1,000 feet per minute, according to the degree of fineness required; the higher the speed, the coarser the product will be. The size of the feed to a large extent decides the diameter of the rolls, but in any case to get the best results from a machine the material to be treated should not be more than four or five times as large as the product required. High speeds can be best obtained when each roll is separately belt-driven, which method also allows a certain amount of slip to occur, thus avoiding excessive grinding of the roll face. Where the rolls are geared together both a lower speed and lower efficiency are the result. The diameter of the rolls varies from 8 to 36 inches, but a very common size in practice is 18 inches, and, it is worth remarking again, it is better to reduce by steps than in one operation, for not only is the first cost often less, but also the running costs are most decidedly so.

The capacity of the rolls depends to a great extent upon the nature of the material to be crushed, and to a lesser amount upon the width of the rolls, which are made 10 to 42 inches, according to the diameter chosen; but the narrow roll has the advantage in that it can be run at much higher speeds, and for this and other reasons the 12-inch roll is in common use.

An important factor in the efficiency and economy of upkeep of this machine is the provision of a method of even feeding, so that even wear on the faces of the rollers results. Crushing rolls are usually of heavy construction and made in all sizes, having a capacity from 5 to 65 tons per hour and requiring from 5 to 100 horse-power.

Fig. 3 shows the general arrangement of high-speed crushing rolls as made by Edgar Allen and Co., Ltd., Sheffield.

The frame is in one piece, on which are cast two pedestals for receiving the bearings of the fixed roll. The bearings for the other roll have machined soles which slide in dove-tailed grooves and are held in position by powerful springs attached to forged steel tension rods which pass through the bearings and frame. The roll

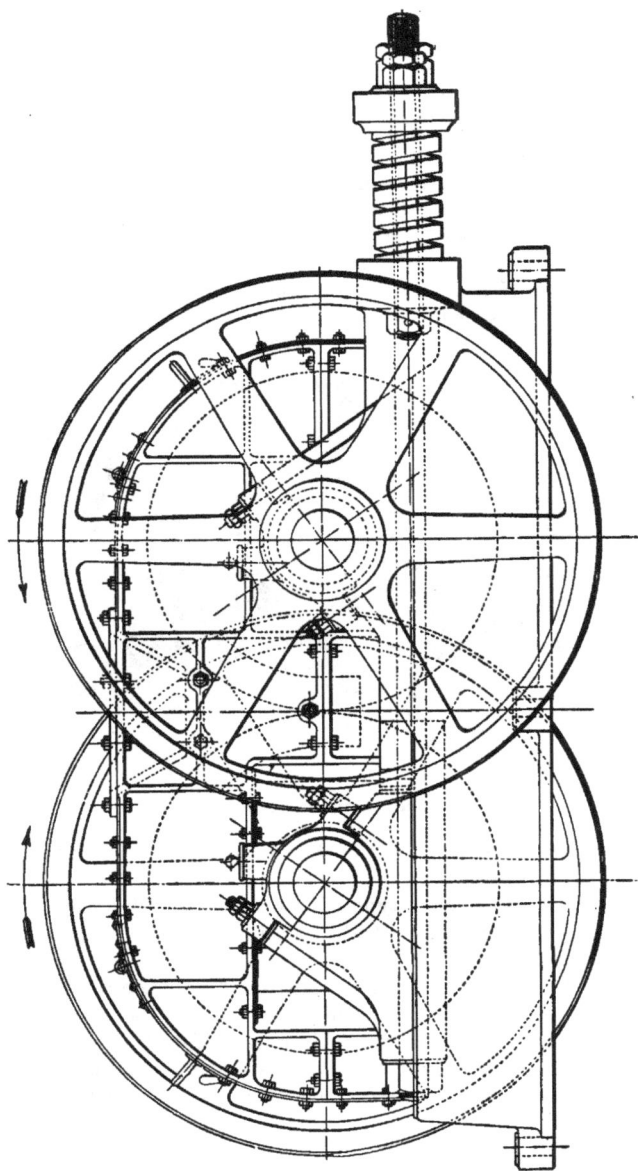

FIG. 3.—G.A. HIGH-SPEED CRUSHING ROLLS

shells are made of cast steel or manganese steel, and are fitted to cast-iron centres in such a way that the shell can be renewed easily when required. Each roll is fitted with a heavy cast-iron belt pulley which also acts as a flywheel.

The following particulars of one size are given as a guide: Diameter of rolls, 18 inches; width of rolls, 12 inches; r.p.m. of rolls and driving pulleys, 100 to 200; diameter of driving pulleys, 36 inches; width of driving pulleys, 6 inches; approximate h.p. required, 5 to 10; approxi-

FIG. 4.—FINE CRUSHING ROLLS, TYPE 5, CLASS 2.

mate product per hour when rolls are set $\frac{1}{4}$ inch apart, 4 to 8 tons.

Fig. 4 shows a type of low-priced machine made by J. Harrison Carter, Ltd., Dunstable, to meet a growing demand.

These machines are constructed in a cast-iron sectional frame which is protected from undue strains by relief springs behind the movable roller. The renewable roll shells are rigidly secured to the shafts by the improved sliding internal cone arrangement; the bearings are of swivel type, instantly renewed or replaced, self-lubricating, and are protected from dust or grit; the rolls are direct driven, each roller by a separate belt. The hopper illustrated contains an automatic feeding device.

Fig. 5 shows a similar type of machine made by the same firm, but having only one roller driven, the other roller following, an arrangement which is supplied when only a single drive is available.

FIG. 5.—FINE CRUSHING ROLLS, TYPE 5, CLASS 1.

Fig. 6 shows a type of machine built by Blair, Campbell and McLean, Glasgow, for sugar-cane work.

This arrangement is suitable where sufficient motive power already exists and can be transmitted to the mill conveniently by belt. The rollers are of special cast metal, having gudgeons of mild steel running in gunmetal bushes, compound spur gearing of special cast iron, with mild steel shafts, and the driving pulleys are of wrought iron for lightness.

Rotary Fine Crusher.—This machine, known also as a "cracker," is almost universally used for reducing rocks of moderate hardness, such as gypsum, limestone, graphite, etc. In its common form it resembles a coffee mill, being in shape like an hourglass and often provided with double doors, so that it can be opened up quite easily and the parts inspected. In its simplest form a shaft with a corrugated-iron shoe revolves within a conical shell having a corrugated inner surface, and by means of an adjusting wheel can be set while the machine

FIG. 6.—SUGAR-CANE MILL.

is running, to give a fine or coarse product as desired. The ordinary reduction is to fragments which will pass a 1½-inch ring. These rotary crushers weigh from 1 to 7 tons, requiring from 1 to 35 horse-power, and having a capacity of from 2 to 30 tons, the largest sizes taking pieces up to 14 inches in diameter.

Fig. 7 gives a sectional view of a rotary crusher.

Edge Runner Mill.—This mill, also known as a "chaser," is particularly useful for dealing with substances such as clay, putty, drugs, chalk seeds, etc., where a very fine reduction is not required. It is also largely used for mixing mortar, for mixing loams for use in the foundry, and for grinding materials which are not of a very hard nature.

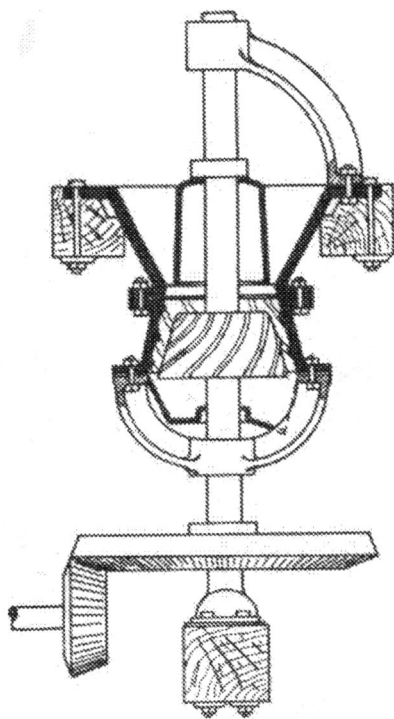

Fig. 7.—Rotary Crusher: Section.

The mill is constructed with a steel or stone bed on which roll two or more heavy rollers of cast iron or stone, known as edge runners or travellers, the whole being contained in a pan provided with a suitable discharging arrangement. In some cases the pan is driven and the rolls rotate of themselves, and in other cases the drive is to the rolls themselves, but in all cases an arm is provided carrying a scraper which travels just in front of the rolls, and which brings the material into the track of the rollers.

As a general rule this machine is used for dealing with batches of material necessitating frequent charging and discharging; but if required for continuous service, grates

FIG. 8.—IRON EDGE RUNNER MILL.

are provided under the rolls having a definite mesh, according to the nature of the material dealt with, and the ground material is gradually forced through the openings into the stationary pan beneath. In some classes of work the rolls are supported at a definite distance above the surface of the pan, so that a certain depth of material is necessary before crushing takes place.

Fig. 8 shows a "Standard" iron edge runner mill made

by Follows and Bates, Ltd., Manchester, which is suitable for pulverizing a great variety of materials, wet or dry, or in oil or water, etc. It is strong, durable, and designed to deal with small requirements. In order to ensure good work, the pan and runners are turned dead true on the faces, and the guides and scrapers are specially designed so as to bring the whole of the materials whilst being crushed directly under the track of the runners.

Fig. 9.—Granite Edge Runner Mill.

When the grinding operation has been completed, the contents of the pan are quickly and automatically discharged by simply turning the handwheel shown in front.

Fig. 9 shows a granite edge runner mill suitable for grinding and mixing crystals, powders, pastes, drugs, etc., either wet or dry, and is made by the same firm. The object sought has been to produce a mill that will operate upon materials in the most effective manner without their coming in contact with iron. To this end the solid bed and runners are made of hard, close-grained grey granite, having surfaces perfectly dressed and dead true. The hopper in which the runners rotate is built up of hard sycamore sections, neatly tongued, grooved, and riveted together in such a manner as to prevent shrinkage or leakage and to be as absolutely clean as the granite itself. The scrapers that clear the runners and the scrapers which are plough-shaped and revolve with the cross-head are of lignum vitæ and keep the bed clear, causing all the materials being ground to pass into and under the track of the runners, which possess a rolling as well as a grinding action, thereby, in a short time, reducing the contents of the hopper to one uniform and smooth consistency.

Fig. 10 shows a type of overhead driven mill with revolving pan, having a perforated bottom with a stationary pan underneath. These mills are specially suitable for pulverizing all kinds of dry materials, such as ashes, lime, gypsum, plaster of Paris, silica, sandstone, bricks for making cement, fireclay, coke, etc., and are provided with a set of grates having a mesh down to $\frac{3}{16}$ inch.

These mills weigh from 3 to 11 tons, requiring from 3 to 20 horse-power, and have a capacity of from 8 to 100 cwt. of ordinary burnt limestone per hour.

Disintegrators.—These machines, also known as pulverizing mills, are especially adapted for dealing with substances of a lumpy nature which are neither hard nor gritty, such as gypsum, dry colours, sulphur, borax, starch, bones, etc. The essential features of these machines are as follows: Two circular plates, mounted on a horizontal axis, revolve concentrically in opposite directions; each plate is fitted with one or more circles of short iron bars, rigidly fixed at right angles, and so arranged that they interlock with those of the other plate, thus forming a circular cage made up of two or more concentric circles of short rods.

The material is fed into the centre of the cage, and

drops down upon the first set of bars, which, travelling
at high speed, beat the material to pieces, and at the same

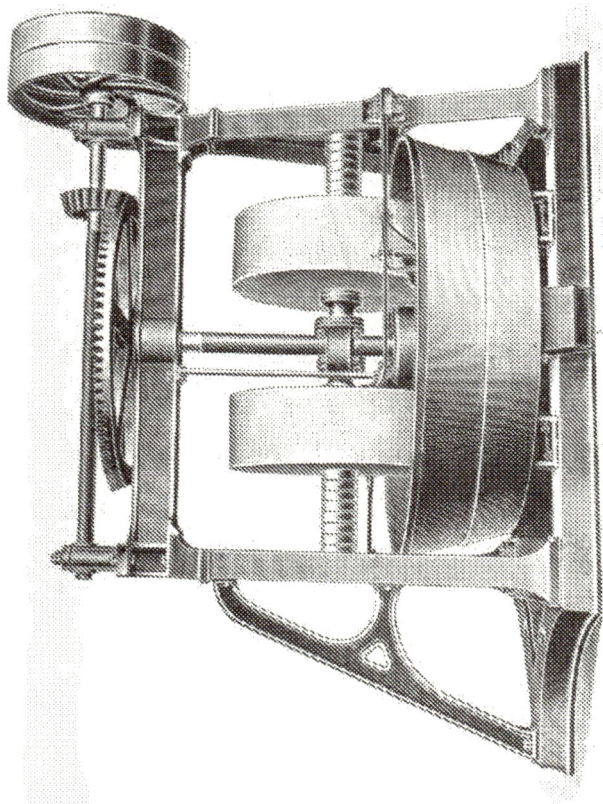

FIG. 10.—MILLS WITH REVOLVING PAN.

time impart a centrifugal motion to it, forcing it through
the bars on to the second set, which are travelling at a
high rate of speed in the opposite direction. By the

time the material reaches the outer bars it has been reduced to a fine powder.

The speed of the machine largely determines the fineness of the product, which is also to a lesser extent affected by the distance apart at which the bars are set. The peripheral speed of these machines often reaches 20,000 feet per minute; they are made in sizes up to 55 inches in diameter, from 3,000 to 15,000 pounds in weight, requiring from 6 to 45 horse-power and having a capacity, which depends upon the degree of fineness sought and the nature of the material, of from 8 to 75 tons in ten hours.

In addition to the usual disadvantages belonging to all high-speed machinery, it is obvious that these machines in running tend to generate a certain amount of heat, which in some cases may become excessive, although the action of the beaters creates a strong air current, which exercises both a cooling and a drying action on the material being pulverized.

These machines must be carefully fitted to a perfectly dust-tight receiving chamber provided with a proper means of discharge, and at the same time the strong current of dust-laden air must be filtered through gauze screens in a dust chamber or allowed to pass to dust balloons, which, being made of porous material, allow the air to escape and the dust to accumulate at the bottom, from which it can be discharged by means of a valve.

It may be again noted that a better result is obtained by allowing the machine to do its own screening and returning the coarser material to the centre of the machine.

Fig. 11 shows a type of disintegrator made by J. Harrison Carter, Ltd., Dunstable, to which the following details refer: The body of the machine is of cast iron, the inside of the circular grinding chamber being lined with renewable chilled-iron sides to reduce the wear. The bottom half of the circumference of this chamber is formed by screens held in position by adjustable screws. A strong spindle carried in self-lubricating bearings runs through the centre of the grinding chamber and carries a cast iron and steel combined disc, which in turn holds a series of from four to six beaters, the tips of which run close to the inner circumference of the grinding chamber, and, when running, cover the whole width of this chamber.

The beaters are mainly responsible for the grinding, being assisted, however, to some considerable extent by

FIG. 11.—DISINTEGRATOR.

the ratcheted sides and top of the machine. The beaters are easily changed and are hardened to resist wear; they can also be cheaply repaired when worn.

The use of screens of various meshes enables the finished material to be ground to any desired size, fine or coarse, and also allows the finished material to escape at once. The mesh of screens, varying as they may from $\frac{1}{64}$ inch and rising in gradations of $\frac{1}{64}$ inch, up to

2

a 3-inch mesh, enables the disintegrator to treat almost every article likely to be ground, with the following exceptions: Substances of a very gritty and cutting nature, such as hard quartz, hard limestone, cement clinker, flint, and similar materials, or those containing a large percentage of moisture, such as plastic clay. The screens and, to an extent, the speed regulate the degree of fineness to which any material is reduced; and these screens, being made in very varied meshes—viz., with from $\frac{1}{16}$-inch to 3-inch spaces—enable a great variety of materials to be ground and a large number of grades produced. The grinding and discharging action is continuous, and to obtain the full output from the machine the feed should also be continuous and even. The speed, to obtain the best results, must be kept regular and maintained at the full value given for the machine.

Many manufacturers object to the use of sifters to follow the grinding machine; this is generally false economy, both as regards output and power. For example a No. 1½ machine fitted with $\frac{1}{64}$-inch screens would grind about 4 cwt. per hour of sugar or similar material. The same machine fitted with a $\frac{1}{16}$-inch screen would pass about 10 cwt. per hour, 60 per cent. of which would be as fine as that passing the $\frac{1}{64}$-inch mesh screens. This 60 per cent. or 6 cwt. per hour would be taken out by the sifter, leaving the 40 per cent. of overtails to go again to the disintegrator to be further reduced, thus showing a gain of 2 cwt. per hour. The sifter and grinder should always be so connected that they work automatically, in order that there should be no additional cost for labour. Further, the material, to pass a disintegrator fitted with a $\frac{1}{64}$-inch screen, must be quite dry, whereas a material containing a fair percentage of moisture would readily pass a $\frac{1}{16}$-inch screen.

Figs. 12 and 13 show sectional views of this type of machine.

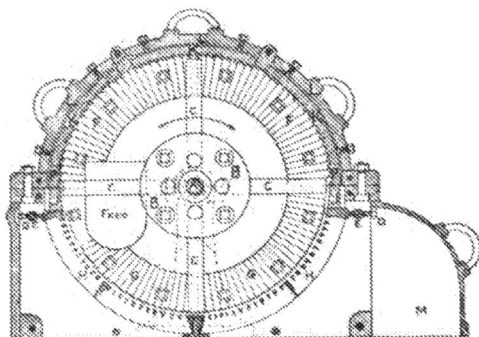

A—Spindle
B—Disc
C—Beaters
D—Screw for adjusting Screens
E—W.I. Cross Bar –
F—Top Side Ratchets
G—Bottom ditto
H—Screens

J—Bottom Screen Block
J—Top Screen Block
K—Top Door Ratchet Linings
L—Top Doors
M—Special End Door
N—Ordinary End Door
P—Top Cross Block for 3½ and
    4½ machines only

Fig. 12.—Disintegrator: Section.

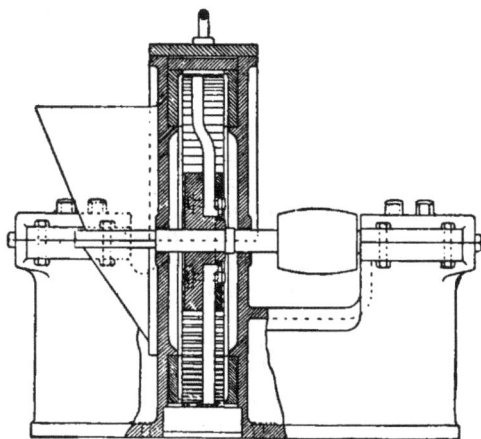

Fig. 13.—Disintegrator: Section.

It is requisite for the efficient working of these machines that a continuous current of air pass through them. This current of air, when the machine is properly fed, enters with the feed and also around the spindle at the centre of the machine, and continues to do so as long as the screens are clear. The air passes, in the ordinary working, with the ground material through the screens into the box or stand, and becomes charged with dust, and as the action of the beaters is somewhat fan-like a considerable air pressure is caused inside the box. This pressure, if not relieved, causes back pressure against the incoming material, and in consequence greatly lessens the grinding capacity, and also causes a greater consumption of power in driving. If air blows out of the centre, it is generally an indication of back pressure or overfeeding. In all cases when the receiving box is made air-tight a trunk is led away from the top of it to a dust chamber or dust balloon. The trunk should be as large as possible, fixed as nearly vertical as it can be, and with as few bends as possible. The dust room should be made as large as possible, so as to allow the air to expand and come to rest, and thus drop the suspended dust.

Figs. 14 and 15 show a standard fixing as supplied by J. Harrison Carter. The disintegrator is fixed upon a dust-tight wooden grinding box or stand placed upon the ground. The machine is driven by a counter-shaft either suspended from the roof or bolted to wooden sleepers in the ground. The counter-shaft is driven off the flywheel or pulley on an engine or motor. In this arrangement the stand shown is made of a top and bottom framework of timber, the timbers of each frame being morticed and securely pinned together. The top and bottom frames are then joined together at the proper distance apart by four vertical legs morticed and bolted to both frames. Two other cross-timbers are also required in the top frame to receive the holding-down bolts of the machine; these

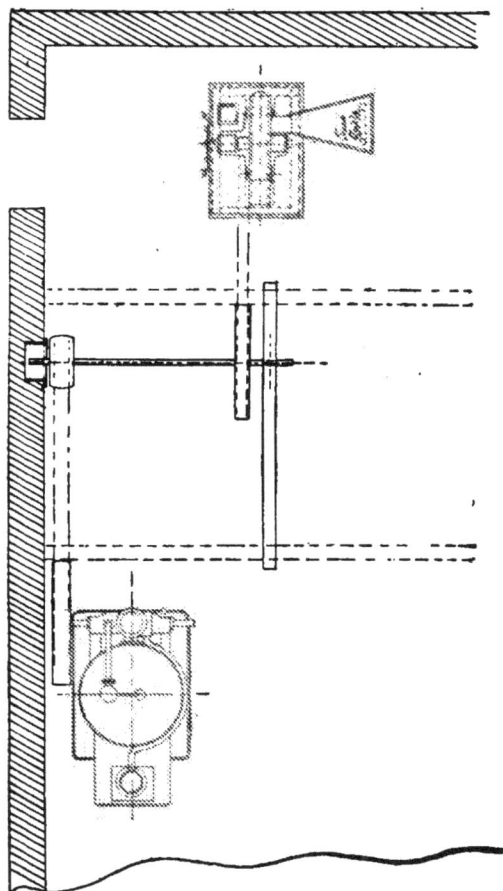

FIG. 14.—G.A. DISINTEGRATOR PLANT: PLAN.

timbers should be placed parallel with the grinding chamber and not across it, so as not to block up the discharge from the machine.   The top of the stand should be covered over with planking, laid down perfectly even and flat, the joints being placed parallel to the cross-timbers, tongued and made dust-tight.   The planking should be securely fixed to the top of the frame and an opening cut between the cross-timbers for the escape of the ground material. The stand should be completely enclosed by match-boarding, made dust-tight, and a sliding door, having a dust-tight seating, provided in one of the sides to allow of periodical cleaning.   It is of the greatest importance for the proper working of the machines that plenty of room is left underneath them.   The stands or hoppers should be made as deep as convenient, so as to leave room for the air blown into them by the revolving of the beaters. In no case must a machine be fixed so that there is not a free escape for this air, and in order to get rid of this air and dust a spout should be taken from the top of the stand and led to a dust room or balloon.

Great care must be taken that the material is fed into the machine as evenly as possible.   The feed should enter continuously and not intermittently, and should be examined for any foreign substance such as iron.   Before starting to feed, the machine should be allowed to attain its full speed judged by the hum of the beaters, and the rate of feeding regulated by this hum.   In very fine grinding, when it is of importance that no pieces of grit should appear in the product, the screens must be packed by a layer of putty or string along that part of the screen which rests on the inner lining of the machine.

The bearings are self-lubricating, and should be inspected two or three times a day and new oil put in every other day.

When the beaters are repaired or replaced, care must be taken that they (the disc and the spindle) are perfectly balanced, so that on turning they show no tendency to come to rest in any particular position.

The disintegrator has been somewhat fully treated owing to the great range of usefulness which it possesses,

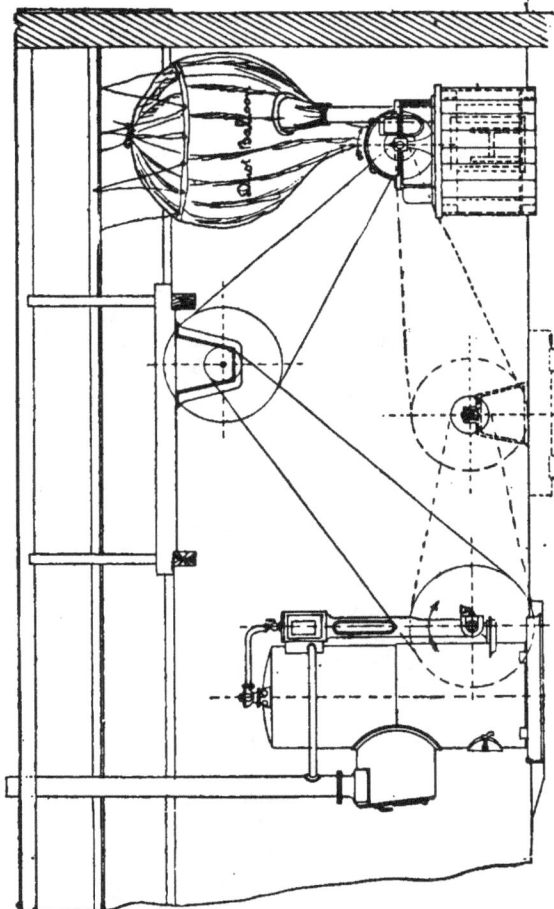

FIG. 15.—G.A. DISINTEGRATOR PLANT: ELEVATION.

as will be seen from the following list of materials which it is claimed are reduced by this means: Alkali, alum, ammonia, anthracene, antimony, asbestos, asphalte, bark, barley, barytes, beans, biscuits, blacklead, blood manure, blue, bones, borax, breeze, bricks, cattle foods, chalk, charcoal, coal, clay, cocoanut husks, copper ore, cork, feathers, felt, fuller's earth, gas carbon, glue, guano, gum, gypsum, hops, iron oxide, lead ore, lime, linseed, magnesite, mica, nuts, oyster shells, paper, phosphates, pitch, rags, resin, salt, shavings, soap powder, sugar, tan, wheat, wood fibre, etc.

Buhrstone Mill.—This machine is in very common use for dry or wet fine grinding of materials such as flour, steatite, graphite, pigments, dry colours, etc.

A buhrstone mill consists of two rough, siliceous discs, one of which is stationary and the other revolving against it. The stones are usually of French buhr or Derbyshire Peak, set either horizontally or vertically. Radiating grooves, about $\frac{1}{2}$ an inch deep and 2 inches wide, inclined to the radii, are cut into the grinding surface of each stone. These grooves must be recut every time the stones wear smooth, which in practice is about once a fortnight, extra stones being provided so that no time is lost during redressing. As the stones naturally wear more towards the outer edges they are usually dressed slightly concave. A much more satisfactory stone is one built up upon a centre of buhrstone with concentric rings of grooved emery blocks, the whole being surrounded with an iron band and having the loose blocks cast in metal.

This type of stone was devised to remedy the wearing away of the outer part of the buhrstone more rapidly than the centre, and in practice not only requires less frequent dressing, but also can be safely driven at 5,000 feet per minute peripheral speed. Usually the material is fed through a hole about 10 inches in diameter in the upper stone, being driven by centrifugal force to the outer edge and cut by the sharp edges of the grooves. To secure an efficient and sweet running mill the rotating stone must be carefully balanced, and if this is done a speed of 4,000 feet per minute instead of 1,000 may be attained with safety.   The limit of speed and therefore of output, apart from the mechanical strength of the stones, is determined by the amount of heat evolved which may be deleterious to the material ground, and to prevent any excessive amount of which a water-cooling arrangement is often provided.

FIG. 16.—MILLSTONE MILL.

The degree of fineness of the product is regulated by means of set screws at the side, and for paste or paint grinding a scraper is provided on the rotating stone to remove the product as it passes the grinding face.

The vertical mills grind faster but not as uniformly fine as the horizontal mills.   In some cases the rotating stone is of smaller diameter than the stationary stone, and is driven eccentrically to it, thereby redressing one another at the same time. Such mills are particularly useful for wet grinding and for paints, graphite, chalk, etc.

Fig. 16 shows a millstone mill made by J. Harrison Carter, Dunstable, in which the stones are made of French buhrstone or Derbyshire Peak, according to the material to be ground. If used for grinding cement, phosphates, etc., the mill is specially constructed and the stones thickened.

A very useful modification of the eccentric mill is the **Vertical Runner Mill,** which is mainly used for small

FIG. 17.—VERTICAL RUNNER MILL.

outputs. Fig. 17 shows such a mill as made by W. M. Fuller, junr., Birmingham. The stones are replaced by a mortar and pestle or runner, the latter having about half the diameter of the former. They are made in cast iron, stone, or wedgwood ware, and scrapers are provided for both runner and mortar. In the smaller sizes the runner may be swung clear and the mortar removed for emptying.

Their efficiency is not high, but they are extremely useful for small operations.

Fig. 18 shows a larger size of runner mill made by the same firm. This machine has been specially designed for mixing and grinding chemicals, drugs, fine colours, dyes, inks, explosives, etc. The mortar and runner are entirely enclosed by a cover, which is easily removable, and an automatic discharge is provided which can be used without stopping the machine.

Roller Mills.—This machine combines a very high efficiency with the production of the finest output. It consists essentially of a roll of granite or steel which revolves in a cavity formed by a piece of the same

FIG. 18.—ENCLOSED END RUNNER MILL.

material as the roll, and which is capable of a lateral movement for the purpose of equalizing wear.

The type of roller mill most commonly found in use consists of three rolls driven at speeds varying in a fixed ratio. The fastest-running roll, which is the delivery roll, is fitted with a scraper, commonly referred to as the "doctor," upon the setting and adjustment of which much of the successful working of the machine depends.

Much that has been said previously in connection with the working of high-speed fine crushing rolls applies with equal force to roll mills. To obtain the best grinding action the surface of the rolls must not be allowed to become polished, but must be roughened, either by sand blasting in the case of steel rolls, or by the use of the diamond turning tool in the case of granite rolls. It need be hardly mentioned that it is of extreme importance that the rolls and spindles must be ground and turned dead true, and any wear due to working immediately remedied.

As a necessary consequence the frame and bearings must be of substantial design, and the pressure on the rolls must be capable of being evenly distributed and adjusted to suit the class of material dealt with. The

FIG. 19.—TRIPLE GRANITE ROLLER MILL.

usual practice to secure even wear in the case of triple roller mills is for the central roller to be given a uniform lateral to-and-fro movement of about $\frac{1}{2}$ inch.

Owing to the fine work required, the type of gearing employed is of the greatest importance to secure smooth and efficient running. In a good-class machine the gearing is either plain machine-cut in cast iron or cast steel, or, better, machine-cut noiseless train gear running in oil, with the pinions so arranged as to be capable of adjustment to compensate for any reduction in the diameter of the rolls.

Fig. 19 shows a triple roller mill made by Follows and Bate, Manchester, for amalgamating and finishing white lead, zinc white, oxides, paints, etc. Other forms are made by the same firm for treating ochres, blues, ink, fine colours, enamel paste, etc. The rollers are of Scotch grey granite, fine-grain hard granite, porphyry, cast iron, or chilled iron, according to the class of work to be undertaken, and range in size from 24 inches by 12 inches to 30 inches by 15 inches, requiring from 3 to 5 horse-power, with an output ranging up to 8 tons per day. These machines have a heavy framework, to ensure absolute steadiness, and accurately machined bearing-ways perfectly aligned, with steel plates secured by studs and lock nuts. The rolls are of grey granite mounted securely on steel shafts and protected from accident by means of spiral springs, even wear being obtained by securing an even distribution of strains throughout and a uniform lateral movement of the central roller. The scraper is of steel, and has a fine adjustment device; the gears are powerful and almost noiseless, and the counter-shaft is in adjustable plummer blocks supported by a pedestal bracket. Patent parallel roller adjustments are provided to front and back rolls, which prevent " sugar-loafing " and ensure fine work.

Ball Mills.—During the last few years these machines have increased in favour, owing to their simplicity of construction, ease of working, low running costs, and freedom from breakdown. They consist of one or more jars of iron or stoneware arranged horizontally in a frame and rotated about a common axis. The grinding action is produced by means of a number of balls or pebbles of porcelain or flint, the jar being driven at a speed of about 80 r.p.m. By using multiple jar machines small quantities of different materials may be treated at the same time, and owing to the small cost of jars the losses due to cleaning may be avoided by keeping separate jars for different materials.

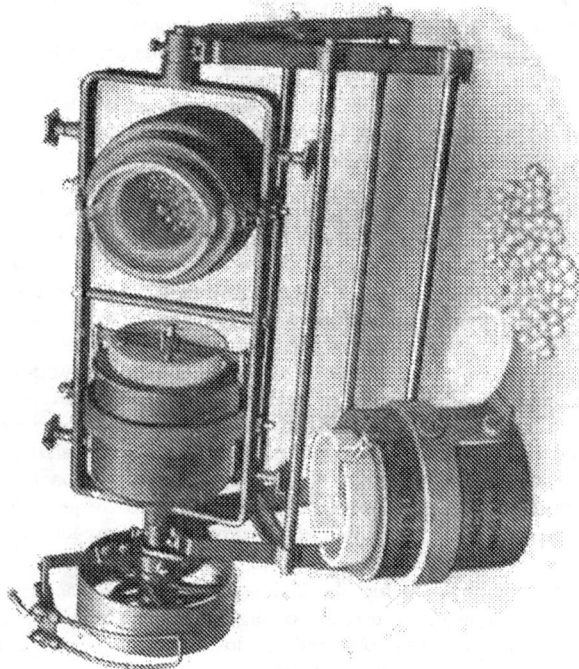

FIG. 20.—BALL GRINDING MILL.

Fig. 20 shows one of the many types of ball mills made by Hind and Lund, Ltd., Preston. These machines create no dust during working, resulting in no loss of material, and pulverize all the material to a uniform product. The capacity of such mills is given in pounds of sand from which the capacity for other materials may be estimated, taking as a basis that 1 cubic foot of sand weighs 90 pounds.

The machine in question is fitted with two porcelain jars 13¼ inches diameter by 12 inches inside, mounted in sheet-iron receptacles, and have an inlet or neck of 8½ inches diameter. The approximate power required is 1 b.h.p. when driven at a speed of 40 to 50 r.p.m. The charge of porcelain balls or flint pebbles for each jar amounts to 45 pounds, giving a grinding capacity of 26 pounds of sand per day for dry grinding and 3 gallons per jar for wet grinding.

For heavy work pebble mills are used having a lining of porcelain, silex blocks, chilled iron or steel plates, the grinding medium being hard flints or porcelain balls according to the class of material to be treated. For special work, steel, brass, hardwood, or vulcanite balls are made.

Fig. 21 shows the " Atlas " pebble grinding mill made by the same firm. The body of the mill is built up on steel gudgeons of substantial design on which the cast-iron side plates are keyed, the outer shell being built up of mild steel plates. Spur wheels are fitted, also barring gear to the end of the counter-shaft, thus allowing the manhole to be brought into position when it is necessary to change the covers for charging or discharging the mill. The gudgeons of these machines are sometimes fitted with stuffing glands and pipe connections for steam or air inlet on one gudgeon and an outlet with cock on the other, to enable grinding to be done under pressure.

The operation of cleaning out the mill is performed by placing in the bottom of the mill a charge of dry sand equal to the given capacity of the mill, and then filling in carefully by hand the charge of balls or flint pebbles. The charge of sand is sufficient to fill in the crevices between the balls or pebbles, and the total volume is equal to half the volume of the mill. The manhole is then closed by the solid cover, and the mill rotated at the specified speed for several hours. The solid cover is then replaced by a perforated cover, and the mill again run until all the sand is sifted out, only the balls or pebbles being retained.

It is of the greatest importance that pebbles should be of the best quality, as soft pebbles not only wear out rapidly, but also deteriorate the quality of the material

FIG. 21.—" ATLAS " PEBBLE GRINDING MILL.

being pulverized. Pebbles of uniform shape, round or oval, are preferable to those of irregular shape, and greatly increase the grinding capacity. On no account should chips or fragments of pebbles be allowed to remain in the mill, as they lower the efficiency considerably.

As the main action of these machines is that of grinding

and not crushing, all material must be crushed to a suitable degree of fineness before being fed to the machine.

Fig. 22 shows a form of ball mill made by Edgar Allen and Co., Ltd., Sheffield, and designed for continuous working. The periphery of the mill is made of hard steel

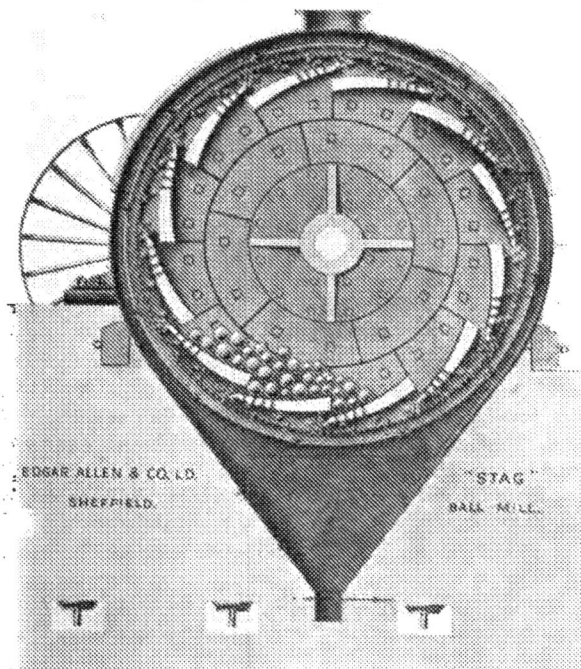

FIG. 22.—" STAG " BALL MILL: SECTION.

grinding plates, stepped as shown; the plates, being perforated, allow the material to leave the inner chamber of the mill as it is reduced to powder; that portion passing from the inner chamber falls on to a second per-forated plate or check sieve, which allows only the finer

3

portion to enter the outer chamber, on which is fixed a final series of sieves, so arranged as to produce the necessary fineness. In each case the rejected portions are returned automatically to the inner chamber for further reduction; consequently, the process of grinding becomes continuous and automatic. The ground material is delivered from the bottom or hopper portion of the chamber into bags by operating a slide, or the bottom may be left open for the finished material to be carried away by a conveyer. The usual type of machine can be fed with material up to 2 inches cube, but machines are made capable of taking material up to 7 inches cube.

The side plates, which are of rolled steel in the larger sizes and of cast iron in the smaller sizes, are mounted on cast-iron centres keyed on to the main shaft. The feed hopper, which is bolted to the inner edge of the main sole plate, is of heavy construction, so as to remain steady under all conditions of working.

The dust casing, constructed of steel plates in sections, with angle iron joints, consists of two parts, of which the top one is fitted with a nozzle from which the dust generated by the rotary action of the mill may be carried away to a balloon or dust settler. Where only limited power is available, a friction clutch is substituted for the fast and loose pulleys, so that starting up may be accomplished more easily and with less shock to the gearing.

These mills require from 3 to 60 b.h.p., and are charged with from 3 to 60 cwt. of steel balls according to their capacity. The main purpose of these mills is to reduce material to a suitable degree of fineness in order to feed a finishing mill.

Tube Mills.—The tube mill is essentially a machine for fine grinding, and since its introduction a few years ago it has replaced practically all other machines for this purpose. It is essentially a special form of ball mill, as the grinding is effected by the rubbing of the material

between the flint pebbles or steel balls and the sides of
the mill, but a certain amount of crushing is performed
by the rolling of the balls and also by their impact in
falling after they have been raised a certain distance by
the revolution of the mill. The difference in action is
that the material to be ground is fed in at one end and is
delivered as a finished product at the other, the degree
of reduction being controlled by the speed of the feed,
since the longer the pebbles are allowed to operate, or,
in other words, the slower the feed, the finer will be the
condition of the ultimate product. Some machines are
provided with a spiral worm feed whereby a certain amount
of material is allowed to travel to the grinding chamber,
and from whence, after passing a perforated plate, it is
carried by another worm and discharged.

These machines are capable of being used for either
dry or wet grinding, and in the former case the material to
be ground must be quite dry, as 1 per cent. of moisture
will seriously reduce the output, and in the latter case,
for wet grinding, sufficient moisture must be present
so as to form a sludge or slurry.

Edgar Allen and Co., Ltd., Sheffield, divide tube mill
grinding into four classes and provide machines accordingly:

1. For grinding either wet or dry material which has
been previously roughly ground or pulverized in a pre-
paratory mill such as a ball mill (Fig. 23).

2. For the preliminary treatment of either wet or
dry material which has been reduced to a size equal to
about 2 inches cube.

3. For the preliminary and final treatment of either
wet or dry material which has been reduced to a size
equal to about 2 inches cube.

4. For the treatment of dry material, in conjunction
with air separation, the material having be en reduce
to a size equal to about 2 inches cube.

The first type of mill, made in several sizes according

EDGAR ALLEN & CO. LD.
SHEFFIELD.

STAG
TUBE MILL

Fig. 23.— "Stag" Tube Mill: Section.

to the quantity and fineness of the finished product required, is designed to carry a charge of flint pebbles in some cases and small steel balls in others.

In the case of flint pebbles being used, the mill is lined with either quartzite or chilled cast-iron plates, the end lining plates being of manganese steel. The quartzite lining is preferable to cast iron, both on account of its longer life and the fact that the efficiency of the mill is increased, due to the quartzite bricks having a rough face, which prevents slip of the pebbles down the sides of the mill.

The mill carrying a charge of small steel balls is suitable for dealing with refractory material, and is lined with steel plates designed to prevent slip. The fineness of the feed supplied to this machine should be such as to pass a 16-mesh sieve—viz., one which has 256 holes per square inch.

The second type of mill is intended for preliminary reduction only and the product passed to a finishing mill as described above. This mill is lined throughout in hard cast steel or manganese steel, and the diameter is larger in relation to the length as compared with a finishing mill. It takes a feed up to a size equal to a 2-inch cube, and is charged with steel balls from 3 inches to 5 inches diameter.

The third type of mill is a combination tube mill, being divided internally by a special diaphragm into two chambers, one of which contains steel balls, the other, which is the finishing chamber, containing flint pebbles. Hard cast steel is used for lining the first chamber and quartzite for the remainder, so that this machine is well suited for grinding cement, clinker, coal, and various kinds of ore.

The fourth type is practically the same as the second type, but it is used for " bulk-grinding," by which is meant that a large quantity of feed is given to the mill, but only a portion is reduced to the required fineness during its first passage. The whole of the product, coarse and fine, is passed through an air separator, which extracts the fine portion and returns the oversize to the mill for regrinding. The advantage of this system is that there is a saving of power required, and that by means of adjustments made at the air separator the fineness is controlled through a wide range.

FIG. 24.—STAMPS FOR CRUSHING.

These mills are made up to 30 feet in length and 6 feet diameter, carrying a charge of 350 cwt. of pebbles and requiring about 180 h.p.

Stamps.—This type of reduction machine performs its work by the simple method of repeated blows on the material by means of a falling weight under the action of gravity or power. Although of very poor mechanical efficiency, the low running costs make them very suitable in the larger sizes for the mining industry, and in the chemical industry the small sizes are found decidedly useful for reducing material of a sticky or oily nature, and for nuts, mustard, etc.

Fig. 24 shows this latter type of machine, made by J. Harrison Carter, Ltd., Dunstable, consisting of two iron pots having heavy stamps lifted by cams and dropped by their own weight into the pots.

Although there are many points of interest in stamps as used by the mining engineer, it is felt that they are not strictly within the scope of this volume, and for further information the student is referred to books dealing with mining machinery.

CHAPTER II

SEPARATING AND MIXING MACHINERY

In the chemical industry it may be necessary either to separate different sized particles of the same material or particles of different nature from one another, and for each class of work distinctive machinery is used.

The simplest form of sifting machine is known as the "Grizzly," and is used, for the sake of economizing power, for separating the smaller pieces of material from the larger, so that the former can go to the fine crusher direct and not with the latter through the jaw-crusher.

A grizzly is an incline built up of parallel bars set transversely, an inch or more apart, according to requirements.

For finer work the ordinary sieve or screen, in which the screening surface is formed by a plate having slots punched through it or by a woven wire, is in common use.

The obvious development of the common sieve is the cylindrical or conical form, which can be rotated about its axis.

The Trommel.—This machine needs very little description, as it consists of a cylindrical perforated plate mounted so that, in the smallest sizes, it can be rotated on an axle, or, in the larger sizes, on friction rollers. The cylinder is set at a slight inclination so as to pass the material through rapidly, and very often several cylinders having different degrees of perforation are arranged concentrically, thus grading the material into several sizes at one operation.

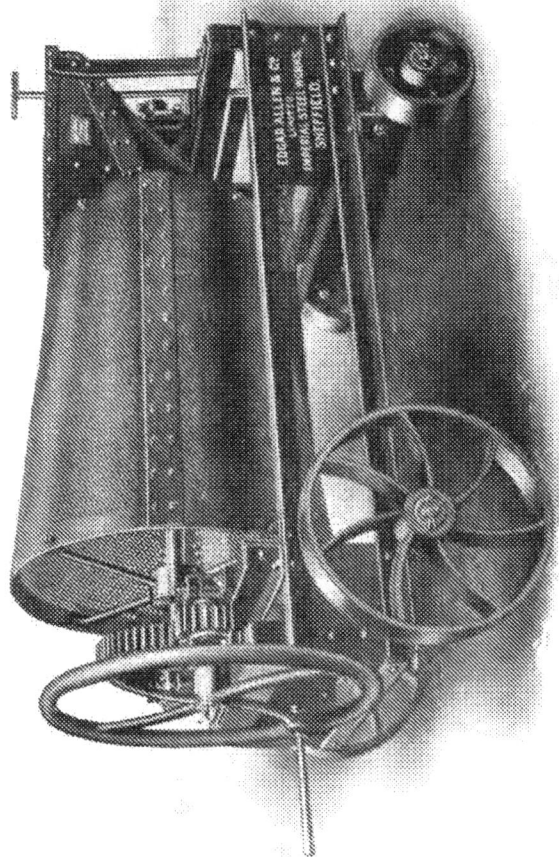

FIG. 25.—PORTABLE SCREEN.

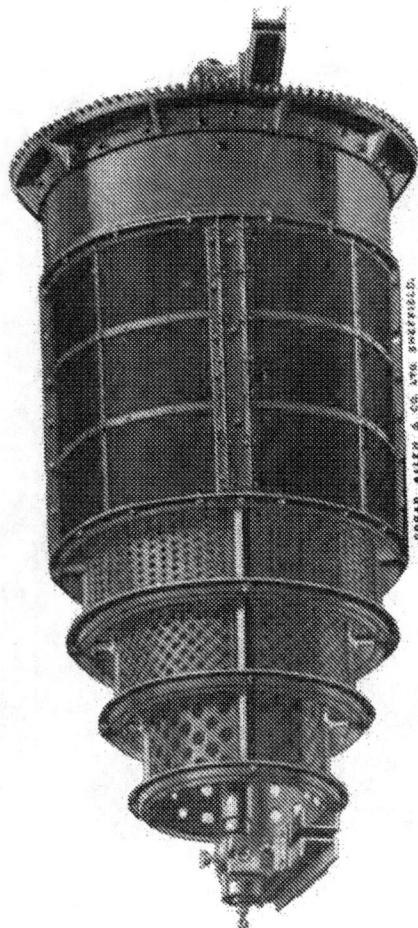

Fig. 26.—Telescopic Revolving Screen

Fig. 25 shows a portable hand-driven screen made by Edgar Allen and Co., Sheffield, and Fig. 26 gives a view of a telescopic screen made by the same firm.

Sifting Reels.—In cases where it is required to grade any ground material into one or more different sizes after it leaves the disintegrator or other grinding machine, or when it is imperative that the finished material be abso-

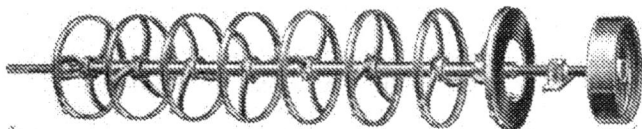

Fig. 27.—Reels.

lutely all of one mesh, or very finely dressed, this machine is the most satisfactory.

For coarse material a cover of coarse mesh is gripped by means of steel straps arranged on a number of cast-iron spiders keyed on a strong spindle, as shown in Fig. 27.

Fig. 28.—Sifting Reels.

To save wear on the cover a wrought-iron ring is fitted at the feed end. These reels are mounted in an inclined position, in casings, which are usually built round them at the factory where installed.

When fine work is desired, the sifting medium is made from fine metal or silk gauze, which requires supporting on either a hexagonal or circular reel. The hexagonal type is suitable in the great majority of cases, but if a very fine product is required and the material is of a sticky nature, the circular reel provided with an exterior brush to keep the cover clean is the best arrangement.

Fig. 28 shows a hexagonal type of sifting reel made by J. Harrison Carter, Ltd., Dunstable.

Fig. 29 shows the " quick-change " powder dresser or sifter for colour manufacturers made by Follows and Bate, Ltd., Manchester. This machine is used for ground colours, ochres, oxides, sugar, flour, blacklead, etc., and

FIG. 29.—POWDER DRESSER.

is so arranged that various grades of powder may be produced at the same time. The provision of a removable barrel allows many powders to be dressed on the same machine, as all parts are designed for quick cleaning.

In the cases of the machines just mentioned a little consideration will show that a large percentage of the sifting medium is inactive, owing to the machine being gravity controlled.

A more highly efficient machine is obtained by the addition of internal beaters or paddles which can be driven at a high speed. By this means the whole of the sifting surface is rendered active, owing to the centrifugal action set up, and the generation of a strong air blast also adds considerably to the output.

Fig. 30 shows a centrifugal dressing machine made by J. Harrison Carter, Ltd., which is manufactured with reels up to 10 feet in length by 2 feet in diameter, driven at 180 to 260 r.p.m. The type illustrated by Fig. 29 attains a size having a reel 20 feet in length by 3½ feet in diameter, but driven only at 40 to 20 r.p.m.

Vibration Machines.—When a charge of material is placed upon a screen it is obvious that the amount of material which passes through depends upon the ratio

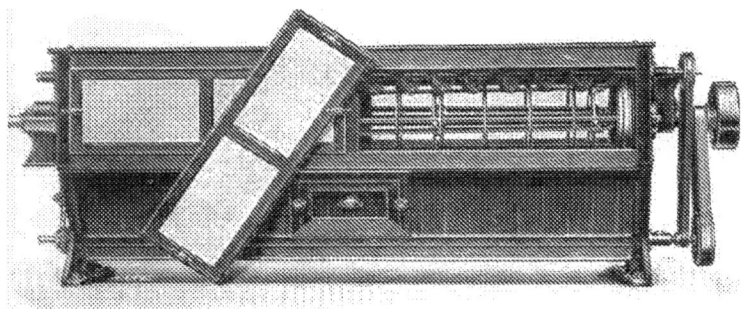

FIG. 30.—CENTRIFUGAL DRESSING MACHINE.

between the sum of the areas of the openings to the total area of the screen. This ratio is known as the opening factor, and varies considerably according to the type of screening surface employed. It is larger for woven screens than for plate screens, but on the score of economy a screen having a long life and a small opening factor is often chosen instead of one with a larger opening factor and of less durable nature. In the case of silk screens, it is essential that the threads should be even and carefully twisted, so that they do not readily become fuzzy, whereby the opening factor is considerably reduced.

If the screen remains at rest it is clear that the bulk of the material remains on the screen supported upon a number of arches formed in the material by the falling away of the portion which has passed through the screen.

Further sifting is only obtained by causing motion of the material relative to the screen, whereby these arches are broken down. The type of motion and the amount of power necessary cannot be estimated satisfactorily, so that with any particular type of machine the best method of obtaining efficiency is to conduct a series of carefully checked trials.

The amount of relative motion required is small; hence sifting machines which have a shaking or vibrating screen form an efficient and important class of sifting machinery.

It should be noted that during the period of vibration not only is the screen sifting the material, but the material is also sifting itself. Under the action of gravity the smaller particles pass between the larger particles, each grade of material acting as a screen for finer grades. As the action proceeds it becomes more effective, until a stage is reached when the material is composed of layers of different fineness, the finest being at the screen and the coarsest at the top. Any further action of the screen will only pass the material through the openings until a size of particle is reached which will just pass. At this point choking becomes serious and the wear is excessive, both of which are to be avoided as far as possible. In addition to this, after the material has once become separated into layers a certain amount of power has been used up in uselessly agitating the coarse material which never passes the screen. Hence, for economical working, separation into layers should be carried out in the first place, and then the coarser layers removed before the finer portions are sifted.

Shaking Sifters.—This type of sifter is very effective for grading most materials, into any sizes and number

of grades. It consists of one or more screens supported
in a frame by flat springs slightly inclined to the vertical
and vibrated by means of a cam or crank having an
adjustable stroke. Owing to the fact that both the
speed and the length of the stroke can be varied as
desired, this machine can be adapted for a wide range
of material. Under the combined action of the crank
and the springs the screen travels on a small arc of a
circle, thus imparting an upward and forward motion
to the material and producing very efficient screening.

FIG. 31.—SHAKING SIFTER.

Fig. 31 shows a type of shaking sifter made by
J. Harrison Carter, Ltd., in different sizes up to 9 feet in
length and 1½ feet in width. For dealing with materials
of a sticky or woolly nature a brush is provided to keep
the screen clean, and where the material has a tendency
to cake, the brush is also made to vibrate. When required, this machine is made with two or more screens
arranged one above the other.

Fig. 32 shows a similar type of machine made by Edgar
Allen and Co., which, however, finds its chief use as a
shaking conveyer. It is simple in design, strongly made,
and so arranged that no violent shocks come on any of

the parts. A special feature of this machine is the connecting rod, which automatically takes up any wear

FIG. 32.—SHAKING SIFTER AND CONVEYOR.

and thus prevents any " knocking " from taking place. The rod is also designed to eliminate any bending action at the fixed end, which is the cause of many breakdowns in this type of machine. This is effected by means of a toggle made of special steel, working on knife edges. The supports, made of strong spring steel, are secured to the bottom of the conveyer by means of pivots working on a fixed spindle.

This machine can be used as a screen, a picking belt, or as a conveyer.

A machine in common use in the gypsum and similar industries, called the " Newaygo screen," consists of a highly inclined screen tightly enveloped in metal sheeting

Fig. 33.—Gravity or Leg Separator.

to prevent escape of dust, and jarred by many small hammers automatically tripped on the upper surface of the cover.

Air Separators.—The separation of materials such as cement, phosphate rock, basic and other slags, coke, ores, etc., which are of a cutting or wearing nature, and therefore not suitable for reels or similar machines, is often carried out by means of an air blast.

If such a material is free from dust a very simple machine, known as a gravity or leg separator, can be used with good results. Fig. 33 is an illustration of this type

4

of machine made by J. Harrison Carter, Ltd., in which the feed is delivered by a feed roll in front of the induced air current when the lighter particles are drawn back and fall into their respective divisions.

The air current is obtained from a fan or from an existing air trunk. This machine can also be used for separating iron from materials before they go to a grinding or other machine. Any number of legs can be placed side by side to deal with various grades of the same material, and combined in one frame complete with the fan.

If, however, the material is dusty, the machine has to be arranged so that the fan circulates the same air and does not exhaust dust-laden air into the atmosphere.

Probably the best arrangement for separating powder of any degree of fineness from dry materials is the "Stag" air separator made by Edgar Allen and Co., Ltd., Sheffield.

Briefly described, this is a self-contained apparatus in which a current of air circulating continuously through a descending stream of ground material separates the finer particles from the coarser, the latter being returned to a pulverizer, mill-stones, or other grinding machinery, to be further reduced. The result is a uniformly fine product which can scarcely be obtained by any other method.

Fig. 34 shows a sectional view of this machine, to which the following details refer: The separator consists of an outer casing of sheet iron A, circular in form, together with an inner casing B, separate from each other, for collecting the fine and coarse materials respectively.

Above the inner casing, and fixed on a vertical spindle, is a fan E, with blades, which, when revolved, induces a current of air. Fixed on the same spindle is a disc, E^1, which spreads the material being treated in a thin stream all round towards a fixed hood directly below the fan. The current induced by the fan passes upward and carries with it the fine particles, which are thrown into the outer casing. The coarse particles, which are too heavy to be lifted by the current of air, fall into the

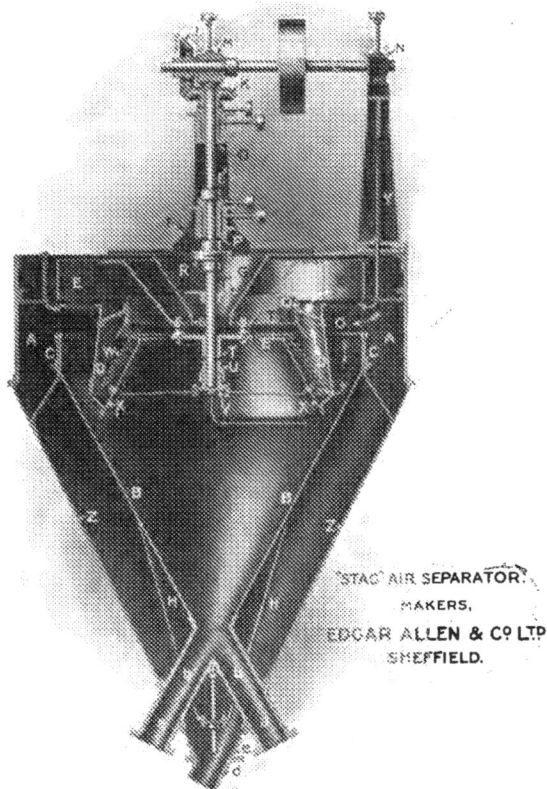

FIG. 34.—" STAG " AIR SEPARATOR.

inner casing, and return by the branch pipes *b* to the grinding machine, to be further reduced. The degree of fineness of the finished material can be regulated by the speed of the fan, also by the partial closing of a damper fixed between the inner and outer casings, which intercepts the current of air.

This machine occupies very little space and requires no settling rooms, and at the same time does away with all brushes, sieves, cloths, etc., in addition to the fact that it can be run at slow speeds of 160 to 260 r.p.m.

It is made in sizes up to 9 feet in diameter and 16 feet 6 inches height over all, requiring about 5 b.h.p.

Among the materials which can be successfully treated in this machine are cement clinker, raw shales, limestone, burnt lime, basic slag, phosphate rocks, blacking, charcoal, gypsum, cattle cake meal, cotton-seed meal, clay and marl, coke dust, chrome ore, bauxite, aluminous earth, fuller's earth, graphite, soda ash, soap powder, gold quartz, etc.

Electro - Magnetic Machines. — Machines in which material is sifted by means of electro-magnetic force are of comparatively recent introduction. The earliest type of machine and one which is in extensive use to-day has for its main object the separation of pieces of iron and steel from material before it is fed into a crusher or grinder, where its presence would cause a breakdown. In its simplest form it consists of a magnet suspended over a belt conveyer or shoot down which the material slides. The pieces of iron and steel caught by the magnet are periodically removed by a workman, who also picks the material as it passes.

A safe and simple type of separator consists of a pair or more of magnetized flat bars which are fixed in the bottom of a spout or carried on a flat table, this table being given a shaking motion or made a fixture. These tables give a large magnetic surface, and consequently the iron has a greater chance of being arrested than it has in the case of some types of barrel and drum separators.

Fig. 35 shows an arrangement for carrying electro-magnets in a spout.

Fig. 36 shows a type of machine made by J. Harrison Carter, Ltd., which is suitable for bone grinders, cattle-food makers, drug grinders, cake grinders, oil mills, etc., where it is essential that no iron or steel shall go to the grinding machinery and that costly hand-picking shall be avoided.

The extraction of the iron is effected by one or more pairs of electro-magnetic bars placed in the bottom of

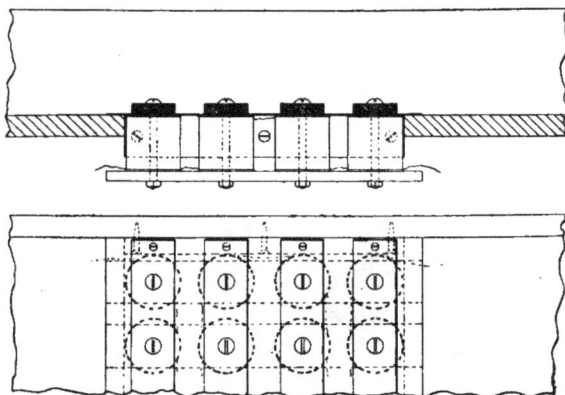

FIG. 35.—ARRANGEMENT OF ELECTRO-MAGNETS IN SPOUT.

a reciprocative tray, which can be made of any desired width and length and fitted with any number of magnets, according to the nature of the material treated or the quantity to be dealt with. An automatic cut-out valve which opens when the current fails acts as a safety device and prevents the feed entering the grinder.

A very simple and effective machine for separating small particles of iron or steel from grain or seed is formed by substituting a specially constructed magnetic pulley for the driving pulley of a belt conveyer.

Fig. 37 shows such an arrangement in action, the grain being shot off by centrifugal force, which in the case of the iron particles is overcome by the magnetic force,

FIG. 36.—ELECTRO-MAGNETIC SEPARATOR.

and causes the particles to be carried round the pulley to a point where they are discharged into a separate receptacle.

The action of the magnetic pulley forms the basis for

several types of complicated separating machines designed for continuous working. By using several magnetic drums rotating at different speeds it is possible

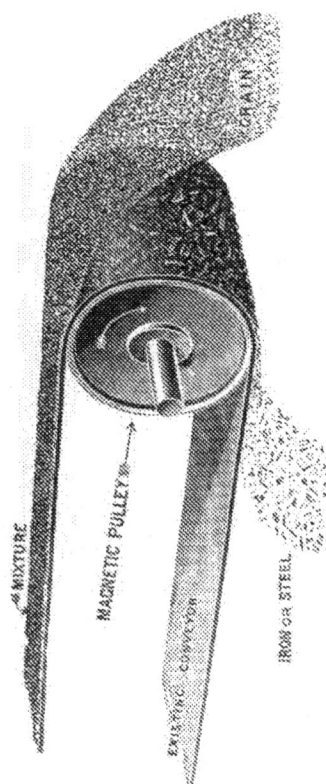

GRAIN

MIXTURE

MAGNETIC PULLEY

EXISTING CONVEYOR

IRON OR STEEL

FIG. 87.—MAGNETIC PULLEY.

to grade material according to the magnetic properties of the products present.

Separating machines which depend for their working upon the different behaviour of materials to charges of static electricity have been developed within recent years. These machines for their successful running require much skilled attention and knowledge of the principles involved, and on that account are not found widely used by the chemical industry of to-day.

Water Separation.—One of the oldest and still most widely used processes for obtaining separation of materials consists of utilizing water as a medium for the separation

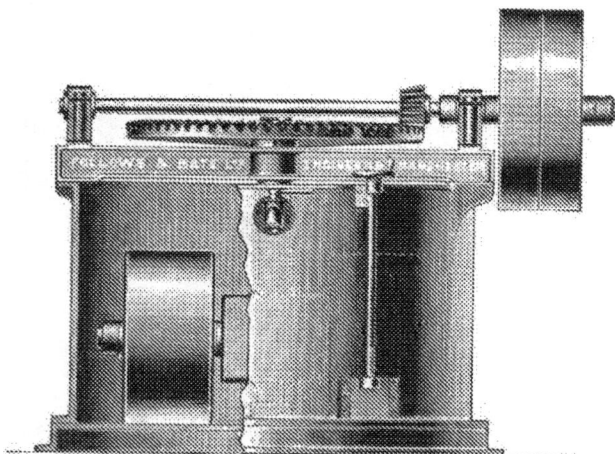

FIG. 38.—LEVIGATING MILL.

of bodies having different specific gravities or of different sized particles having different rates of settlement. In the mining industry this method has received great attention, and the resulting developments have been many, but in the chemical industry it has only limited application.

In the case of barytes, oxides, ochres, coloured earths, and similar substances, the material is first ground under water in a special edge runner or levigating mill, a type of which is shown in Fig. 38, made by Follows and Bate, Ltd.

This powerful machine is designed for crushing hard oxides, etc., in water, and is conveniently arranged for floating off the desired product by means of a tap fixed at the top of the reservoir, inside which the runner rotates. The sludge door is easily opened, by which grit, sand, iron, and other suchlike particles, can be speedily removed.

The mixture from the levigating mill is run into a large vat or buddle, and after standing for some minutes the upper portion, to a definite depth, is run off into settling tanks, where it is allowed to stand for a longer or shorter time, depending on circumstances. After more or less complete settlement the clear supernatant liquid is run off and the sludge removed for drying. Very fine material of low specific gravity may take three or four weeks to settle, but in the majority of cases the operation is completed in a few hours. Sometimes the settling tanks take the form of long troughs, through which the mixture is made to flow at a definite rate. By this means the coarser and heavier particles settle first, the finest being deposited at the outlet, whilst all floating impurities are carried beyond. This method can be usefully adopted for large quantities of material which does not take too long to settle. Many attempts have been made to construct machines which will accelerate the rate of settlement and provide a more compact arrangement, but they can hardly be said to have received much application in the chemical industry.

Fig. 39 gives a view of the interior of a settling tank or subsider for cane juice. It is provided with a copper decanting pipe and float having a limited travel. When the decanting is finished the residue can be run off by means of a separate cock. This type of tank, made by Blair, Campbell and McLean, Ltd., Glasgow, ranges from 200 to 1,000 gallons in capacity, and is often used in a series.

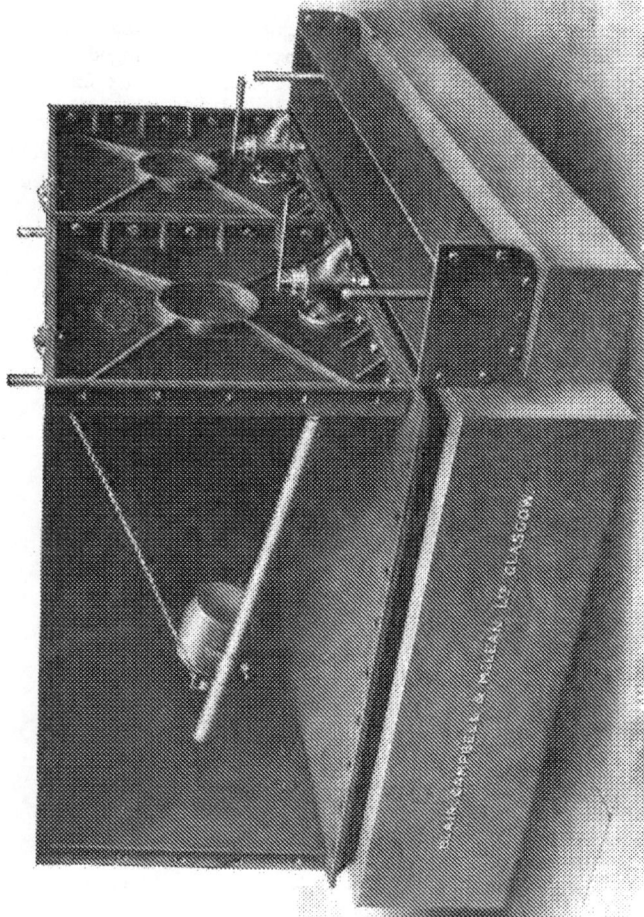

MAIN, CAMPBELL & McLEAN, Lᴅ, GLASGOW

Fɪɢ. 30.—Cᴀɴᴇ-ᴊᴜɪᴄᴇ Sᴜʙsɪᴅᴇʀ

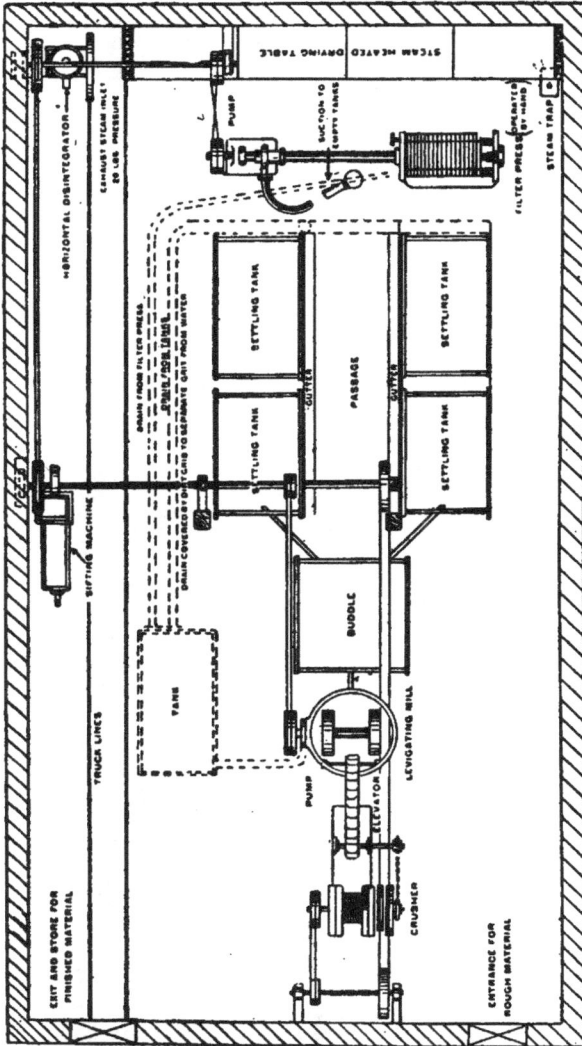

FIG. 40.—DESIGN OF LEVIGATING PLANT: PLAN.

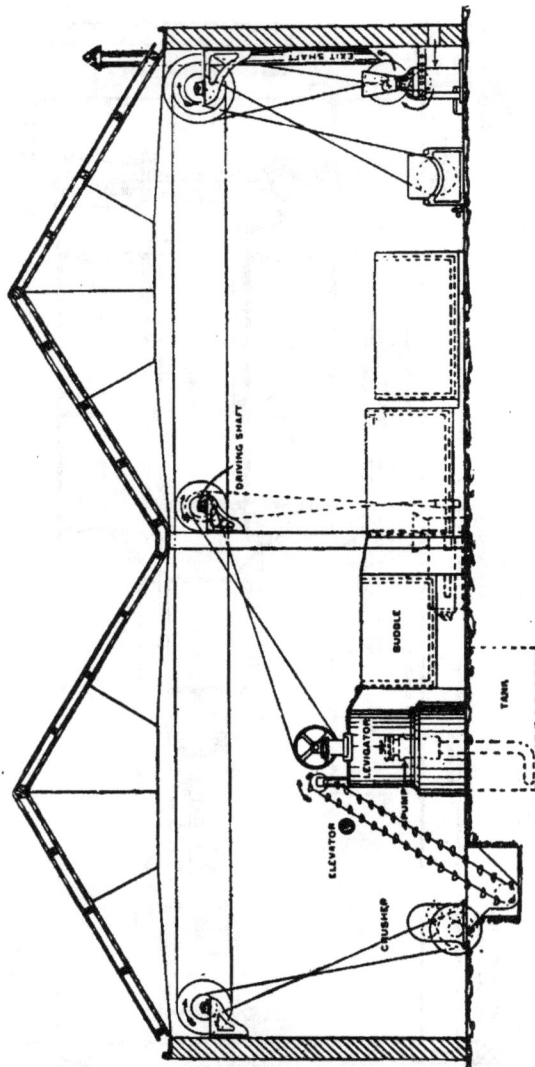

EXIT SHAFT

DRIVING SHAFT

BUDDLE

TANK

LEVIGATOR

PUMP

ELEVATOR

CRUSHER

Fig. 41.—Design of Levigating Plant: Elevation.

Figs. 40 and 41 show in plan and elevation a design for a levigating plant made by Follows and Bate, Ltd., Gorton. A careful study of this design throws considerable light on the number and working positions of the machines which are deemed necessary for the economical working of one of the simplest operations in the chemical industry.

Fig. 42 is interesting as representing the flow sheet of the milling process used in preparing rock salt for market.

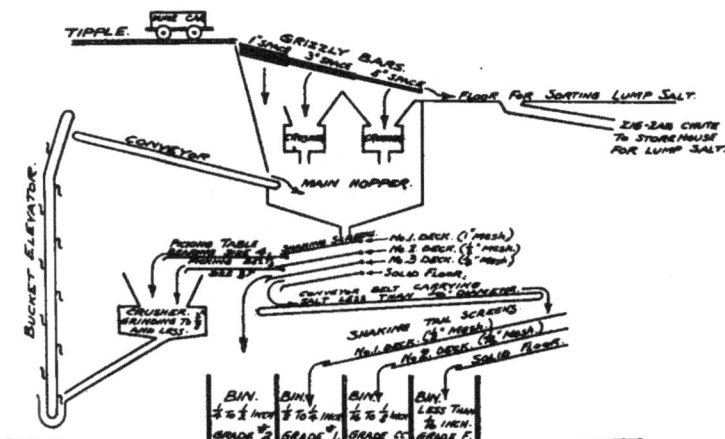

FIG. 42.—FLOW SHEET OF MILLING PROCESS.

Mixing Machinery.—Mixing operations may be roughly divided into two classes—(1) the mixing of solids with solids; (2) the mixing of solids with liquids. To obtain a uniform mixture of different solids is at present a practical impossibility, because as soon as the mixture is set in motion a sifting action takes place, as was explained previously when dealing with the subject of sifting. For this reason such mixing as is required is done at the time of grinding by feeding the different solids, in the required amounts, together into the grinding machine.

The tube mill and the ball mill are commonly used for grinding and mixing at the same time, and when only mixing is desired the balls are removed and projecting arms substituted to assist in turning over the material. This operation is common where different grades of material are required to be mixed, or as a preliminary to the grinding operation, in order to obtain a more uniform product.

FIG. 43.—POSITIVE DRIVEN PUTTY MILL.

The edge runner mill is another machine which is largely used for mixing, with more or less satisfactory results, being limited in the extent of its output.

A special form of this type of machine, made by Follows and Bate, Ltd., is shown in Fig. 43, and is known as a "putty" mill. Besides being useful for crushing to a fine powder, chalk, whiting, chrome, indigo, Prussian blue, etc., it is also handy for mixing into smooth pastes, red and white lead for steam joints, white lead and borings into tough paste, etc. The taper roller and the pan of this machine are caused to revolve by

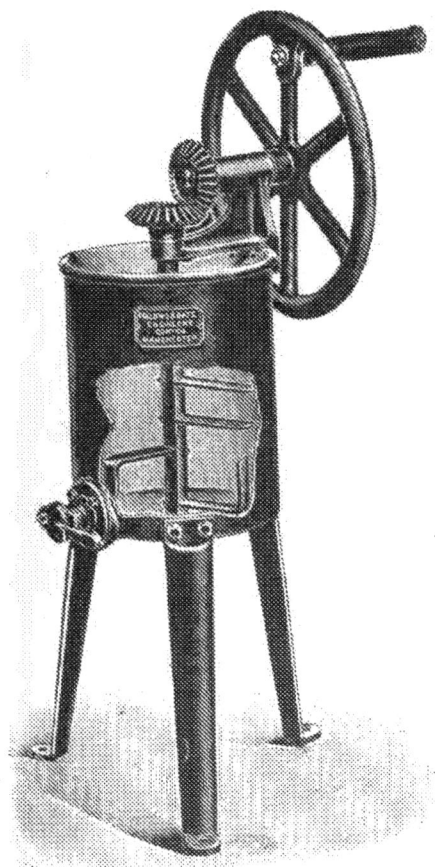

For Hand Power with Fly Wheel.

Fig. 44.—Vertical Pug Mill.

gearing at different speeds, and to act independently of the contents of the pan, no matter how slippery they may be.

For the mixing of solids with liquids to form pastes or semi-liquid products the ball mill is used for high-class work, and the pug mill for paints, enamel varnishes, etc.

FIG. 45.—" POWERFUL " HORIZONTAL MIXER.

A simple form of pug mill made by the above firm is shown in Fig. 44, which gives a view of the interior, showing the rotating arms used for mixing. This machine is useful for small quantities of colours up to 6 gallons, and the hopper is lined with white vitrified enamel to allow of ease in cleaning.

The pugging arrangement is strong enough to mix the stiffest pastes, putty, and the like, and will also mix semi-liquids in varnish or oil with equal facility.

A larger form of this machine is shown in Fig. 45, which is largely used as an amalgamator for ready mixed paints or for oil blending, etc.   It is fitted with a steel pan with double-riveted, lap-jointed vertical seam,

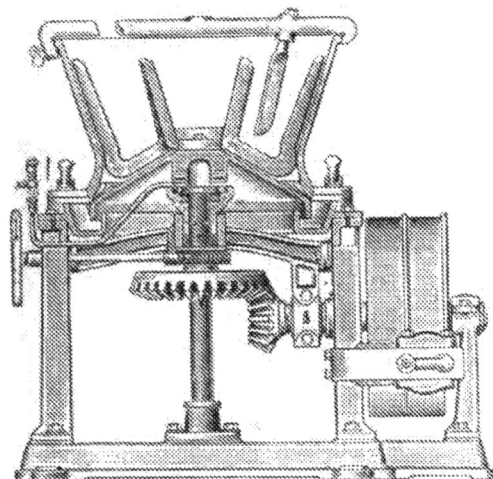

SECTIONAL VIEW OF THE " UNIVERSAL " CONE PAINT MILL.

FIG. 46.—CONE PAINT MILL: SECTION.

single riveted to cast-iron top and bottom rings, a heavy vertical steel shaft carrying forged-steel mixing blades, the angle of which can be adjusted to suit thick or thin material, and a discharge door or cast-iron tap as shown.

A useful and handy type of machine made by this firm for operating upon material of not too great specific gravity is the horizontal mixer shown in Fig. 45. It may be used for the blending of dry colours or powders, or for

5

producing a liquid, semi-liquid, or paste like dough or putty.

The open pan or hopper, of steel, brass, or copper, swings between two heavy standards, and is firmly locked or instantly released by the withdrawal of a stop catch. The toughened steel beaters are detachable for cleaning, and rotate in opposite directions, in such a manner as to prevent the contents of the hopper from gathering in a mass during the process of mixing. The material may be discharged whilst the beaters are revolving by swinging the hopper into any desired position.

FIG. 47.—DOUBLE MIXER FOR SEMI-LIQUIDS.

The "Universal" cone mill made by this firm, of which a sectional view is shown in Fig. 46, is a combined mixing and grinding machine used for paints, enamels, varnish, stains, pulps, grease, lubricants, boot dressing, match composition, antifouling composition, etc.

The cone has deep feeding and fine grooves, so arranged as to force the material inside the hopper outwards to the grinding surfaces, and is balanced on a central vertical steel shaft pivoted at the bottom in a conical hardened steel bearing, being raised or lowered by means of a handwheel at the side. The cone rotates inside an annular trough which is broad and deep, with a square bottom, and is fixed on an incline, so that as grinding proceeds the material flows steadily towards the outlet provided.

For working up thick pastes detachable beaters are provided, and the annular trough is removed, so that the machine delivers on two sides simultaneously. Sometimes it is necessary to keep the materials hot whilst the mixing process is going on, and to this end

FIG. 48.—" OPEN-DRUM " MIXER.

mixing machines are sometimes provided with steam jackets or with an arrangement for heating by gas. Fig. 47 shows a double mixer for semi-liquids or powder provided with a steam jacket, made by J. Harrison Carter, Ltd. This machine can be arranged to work as a charge or continuous mixer with the outlet at the end or anywhere in the length of the bottom.

Fig. 48 shows an " open-drum " batch mixer made by the same firm, which is used for mixing either concrete or tar macadam. This machine is made in sizes weighing up to 2 tons and capable of dealing with up to 20 cubic yards of material per hour.

Fig. 49 shows another machine of this firm's make, which is also useful for concrete and macadam. It is an under-geared mixer which mixes thoroughly without the pieces being crushed smaller than is desired, and from which the material can be automatically discharged

FIG. 49.—UNDER-GEARED MIXER.

when desired. The pan is made up to 7 feet in diameter and revolves up to 19 r.p.m., requiring up to 12 b.h.p.

The soap industry makes great use of mixing machines which are known as "crutchers," from the fact that in the early factories the mixing was done by hand with a wooden stick or crutch.

The crutcher is surrounded by a jacket, into which is introduced either steam for heating or cold water for cooling. There are many different types of beaters, but a common form consists of an Archimedean screw working in a central cylinder, over which the soap passes during the mixing. Various materials, such as borax, starch, carbonate of soda, sodium silicate, talc, sand, perfume, etc., can be added and thoroughly incorporated to produce the many types of soap marketed.

Fig. 50 shows a sectional view of a soap crutcher such as is in common use.

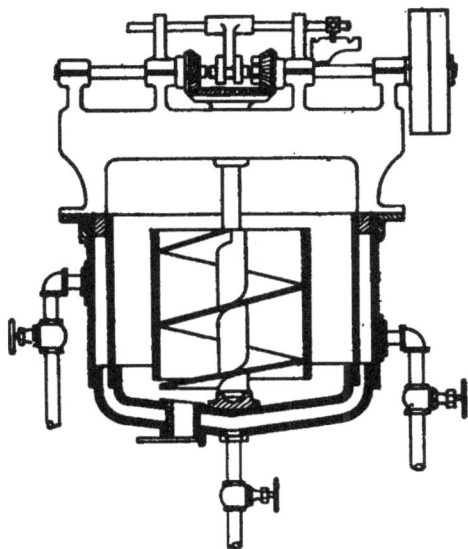

FIG. 50.—CRUTCHING MACHINE: SECTION.

For dealing with plastic materials such as soap, india-rubber, etc., and for the finishing of fine paints, enamels, printers' inks, and suchlike, the fine roller mill is used. In some cases it is necessary to provide the rolls with a steam-heating arrangement, especially where a solvent that has been used is required to be eliminated. These fine rolls have already been treated in a previous chapter, and reference should be made to them in connection with the mixing of materials.

CHAPTER III

FILTERING APPARATUS

FILTRATION is the name given to the process of separating solids from the liquids in which they are suspended, and although a fairly simple operation when conducted cn a laboratory scale, it presents great difficulties when large quantities of material have to be handled. Development has been along the lines of speeding up the process, the actual separation being still effected by the action of some medium such as cloth, paper, asbestos, slag wool, glass wool, unglazed earthenware, sand, or other fine porous material.

The Bag Filter.—Although only capable of dealing with comparatively small quantities of material this type of filter has a wide range of application. As its name implies, it consists essentially of a bag of woven cotton or similar material, into which the material to be filtered is placed, the liquor being allowed to ooze through the pores of the fabric, whilst the solid material is retained.

In dealing with materials of a sticky nature the bag filter has its advantages, and on this account it is found in use particularly in the sugar industry.

Fig. 51 shows an improved type of bag filter for filtering sugar liquors, made by Blair, Campbell and McLean, Ltd., Glasgow.

The top is made loose so that, together with the dirty bags, it can be lifted out of the filter casing and taken by an overhead trolley or crane to a washing tank, and another top ready with clean bags fixed in its place.

This arrangement saves much time and obviates the necessity of having to get inside the filter casing to remove the dirty bags. The above illustration shows a small bag filter containing forty-nine bags. The casing is of wrought iron, and is fitted with a steam coil and mountings, discharge cock, etc.

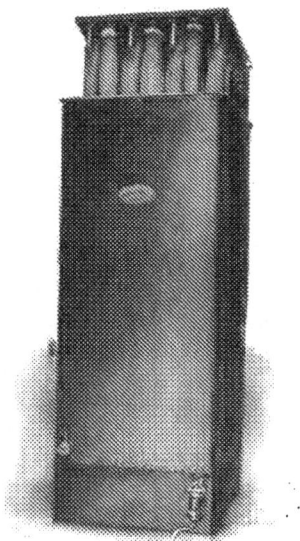

FIG. 51.—BAG FILTER.

The Filter Press.—Filter presses are used in a great variety of industries, and are generally recognized in their present form as providing the most efficient means of carrying out this often-required operation. By the adoption of filter presses the following advantages are obtained: (1) The greatest possible filtering surface is secured, together with the minimum space; (2) a large variety of materials can be treated, as they can be adapted so that the material can be fed into them at pressures ranging from a slight gravitational pressure up to 10 atmospheres; (3) the joints between the filtering plates are under direct observation and control, and access to the internal parts is a simple operation; (4) the solid matter can be washed free of any soluble matter which may be deleterious or may be worth recovering.

There are two principal types of filter presses—viz., the plate and frame type, or frame press, and the recessed plate type, or chamber press.

The **Frame Press** consists of a series of filtering chambers formed by placing alternately a number of solid plates and hollow frames in a suitable framework. The solid plates—which, like the frames, may be constructed of iron, wood, gun-metal, hard lead, aluminium, etc.— have both surfaces corrugated, to allow the filtered liquid to escape easily and at the same time give adequate support to the filtering medium, which usually takes the form of a filter cloth spread over each surface. The rectangular-shaped plate is the one most commonly used, as it is the most economical of filter cloth, but the efficiency of the presses depends upon the nature of the plate surfaces. It is here that manufacturers differ in their construction, each being guided by the results of his own experience.

The Premier Filterpress Co., London, has found, after many years' experience, that vertical corrugations or ribs, with horizontal ribs at the top and bottom, are the most efficient and provide ideal support for the filter cloth, besides helping the filtered liquor to get away easily and quickly.

In this type of filter plate the ribs are quite smooth, so that they can be easily cleaned, and are deep enough to prevent the cloth sagging to the bottom of the corrugations and so prevent filtration. It has been found by experience that a quarter of an inch is a minimum satisfactory depth to which no filter cloth under pressure can penetrate. Should the fibre of the filter cloth be weakened under the action of the material filtered, a perforated sheet is supplied to cover the ribs and give additional support, whereby the life of the cloth is considerably lengthened.

When this type of press is closed up there is formed a series of hollow chambers, each of which forms in itself a complete filtering chamber. Since all the chambers of a filter press commence working simultaneously, it is immaterial how many chambers are employed, except, of course, as regards the quantities of materials to be dealt with, so that, provided the means of feeding the presses are suitable and adequate, a press with a filtering surface of 1,000 square feet will fill in the same time as one having only 100 square feet of filtering surface.

As a general rule solid matter does not form in a cake gradually built up from the bottom of the press, but forms on the surface of each plate and gradually builds up towards the centre, finally forming a complete cake.

Where cakes are required having a thickness of over 1½ inches or require a thorough washing, the frame press is the more suitable type to employ. A further advantage is secured by the ease with which the filter cloth is fixed, as all that is necessary is to cut a piece of filter cloth rather more than double the length of the plate and simply hang it over. This is possible owing to the feed passages, etc., being arranged on the border of the plates with ports leading to the interior of the chambers. Also, in certain cases the frame and cake can be removed and stored without breaking up.

It is often found necessary to wash the cake when formed, either to remove deleterious substances or to recover valuable soluble matter. For this purpose both plates and frames are arranged with channels for feed inlet, wash-water inlet and outlet, and a separate outlet —or outlet taps—on each plate, for the filtrate. The usual arrangement is for the channel for the wash-water inlet to be made at the bottom of the plates and frames; the three channels for feed inlet, wash-water and air outlets at the top; and the filtrate outlet at the bottom corner, opposite the wash-water inlet.

The material to be filtered enters the chamber by means of a port in each frame from the feed inlet A (see Fig. 52). The wash-water inlets and outlets, also the air outlets, are arranged so that the port to the chamber is made only in alternate plates. By this arrangement the water enters at B behind the cloth on one side of each cake, and as it rises in the press expels any enclosed air, which can escape through the air outlet C.

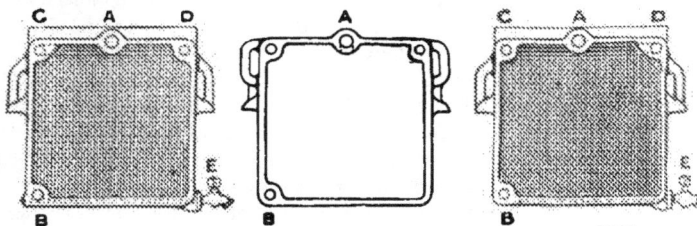

FIG. 52.—FILTER PRESS PLATES AND FRAMES.

The water forms in a vertical wall behind the filter cloth, and passes evenly through the cake and away by the special outlet channel E, on the opposite side of the cake to which it enters. This action is sometimes assisted by putting a siphon pipe on the wash-water outlet, so that even at the top of the press the wash-water is under a slight head.

A special attachment or control apparatus is often placed on the wash-water outlet, in order that hydrometer readings can be taken of the specific gravity of the wash water, whereby the degree of washing can be checked.

Partial washing may be done in any filter press by passing water through the feed pipe, but the above arrangement of a frame press is the only method for thorough washing. With this type a more even thickness of cake is obtained and the feed passage being arranged outside the actual cake, avoids the formation of a core of material throughout the length of the press, which is not subjected to the action of the wash-water.

The position of the feed channel is varied according to circumstances. When the solid matter is so heavy that the formation of the cake becomes abnormal, the feed passage is placed at the top; whereas if the solid matter is very fine and will not form a complete cake the feed is placed at the bottom, to allow of drainage before opening the press. For certain materials which require to be pressed at temperatures above or below the normal, plates are fitted having coils cast internally, through which steam or brine may be circulated at will.

Sometimes steam is admitted into the material itself by means of the cock on the feed inlet or the wash-water inlet. In other cases, in order to produce a drying effect, these passages are used for leading hot or cold air through the cakes when they are formed. As a general rule it is best to have a tap on each chamber, because it gives control, so that if the cloth bursts the trouble can be easily located.

Fig. 53 gives a diagrammatic view of a frame press made by Blair, Campbell and McLean, Ltd., for use in the sugar industry.

Fig. 54 shows a plate and frame type of filter press made by Manlove, Alliott and Co., Ltd., Nottingham.

The **Chamber Press,** or recessed plate type, has plates which are made with raised edges, so that when they are placed together in a horizontal series each pair encloses a chamber, the feed passage as a rule being in the centre. This type is more suitable when materials are used which are liable to clog the passages of a frame press; moreover, when the press is opened the cakes can easily be made to fall out on to a conveyor, truck, or other arrangement underneath the press.

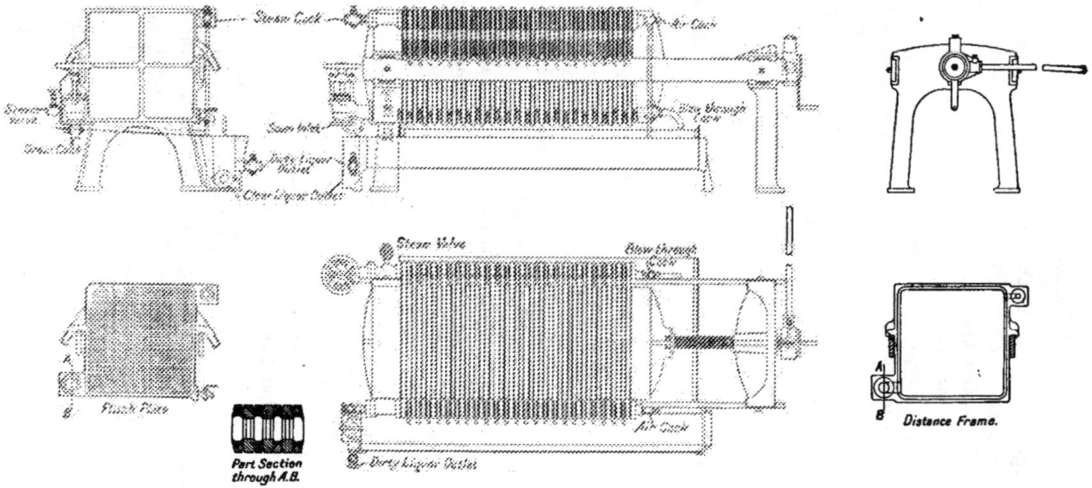

Fig. 53.—Filter Press: Sections.

Fig. 55 shows a recessed type of plate made by Manlove, Alliott and Co., Ltd., Nottingham. The fixing of filter cloths in this type of press is obviously a more difficult

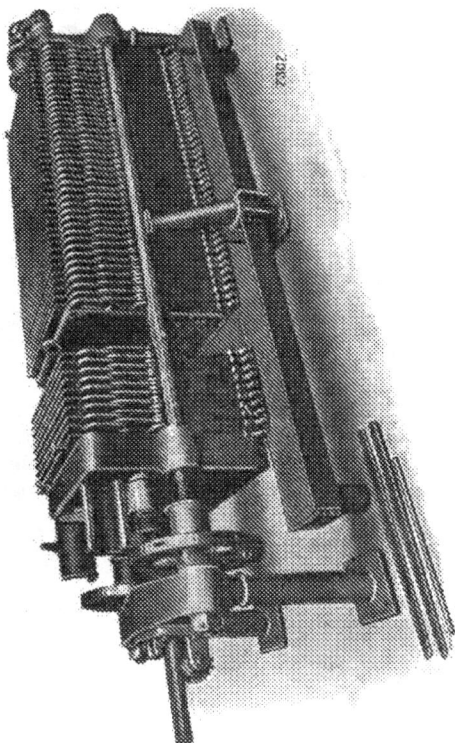

FIG. 54.—FILTER PRESS: PLATE AND FRAME TYPE.

operation than is the case in a frame press. As before, the cloth is cut in pieces rather more than double the length of the plate, but holes must be cut to correspond to the feed channel and the cloth fixed at this point.

This can be effected by means of clips of the " bayonet " or " screw union " type, as made by the above firm and shown in Figs. 56 and 57.

FIG. 55.—FILTER PRESS PLATE: RECESSED TYPE.

Another method frequently employed is known as the " double cloth " system, which employs two cloths sewn together where the feed-hole comes, one half being rolled up and passed through the centre hole and the two corres-

FIG. 56.—FILTER CLOTH CLIPS: BAYONET TYPE.

ponding halves tied together by tapes at the top of the plate. This process can be easily followed out by a reference to Fig. 58. In this type of press the cloth forms an efficient joint between the plates, which grip it between their edges, whereas the frame press necessitates cuffs being slipped over the lugs, or grooves cut round the channel holes and india-rubber washers employed.

Fig. 59 shows a recessed type of filter press made by the above firm.

FIG. 57.—FILTER CLOTH CLIPS: SCREW TYPE.

The materials used in the construction of a filter press depend upon the nature of the materials to be filtered, but where possible iron is used, on account of its greater strength and durability. Wooden presses are made equally as strong as iron ones, but they wear out more

FIG. 58.—FILTER CLOTH: FIXING IN RECESSED TYPE.

quickly; they can, however, be easily replaced at a small cost.

A quick and efficient method of closing a filter press is one of the most important points in their design. There are various methods of closing the presses and keeping them tight during filtration and washing, but experience

FIG. 59.—FILTER PRESS: RECESSED TYPE

FIG. 60.—FILTER PRESS: CENTRAL SCREW CLOSING.

has shown that for presses up to 25 inches square a central screw and handwheel, as shown in Fig. 60, with a lever or capstan bar for tightening up, is an effective arrangement and the least likely to get out of order. To avoid the long operation necessitated by a fixed screw centre, the Premier Filterpress Co., Ltd., provide a rotary screw which after a few turns can be swivelled into any desired position.

FIG. 61.—FILTER PRESS: RACK AND PINION CLOSING

For presses having plates larger than 25 inches square a single screw is hardly powerful enough, and it is here that manufacturers differ in their designs.

The standard arrangement of Manlove, Alliott and Co. is by rack and pinion, with wheels operated by levers for tightening up, as shown in Fig. 61, the rack being connected to the loose head by a flexible joint, by this arrangement considerable simplicity and speed of operation being obtained.

Other types of closing gear, including pneumatic or hydraulic means, as shown in Figs. 62 and 63, are employed in certain circumstances.

The feeding of filter presses is most important, as in order to produce the best results the flow of filtrate must be as uniform as possible.   As the operation proceeds, the resistance, owing to the formation of the cake, increases: hence the pressure rises.   This rise should be slow and regular, and should not exceed a definite pressure, depending on the nature of the machine.   The usual method of feeding is either by a pump, which may be belt, motor, or steam driven, or by means of a forcing ram worked by compressed air.   If pumps are used, those

FIG. 62.— FILTER PRESS: COMPRESSED AIR CLOSING.

having either ball valves or wing rotating valves are most suitable; and whereas for small quantities a double-acting single-plunger pump is good enough, yet for large quantities the three-throw pump, which gives a regular flow, is the one to be adopted.

A more expensive method of feeding is by Montejus and air compressor, but it is not surpassed by any other method for steadiness of pressure.   When using a Montejus the gauge will rise quite regularly all the way through, and if an air receiver is used between the Montejus and the compressor a sudden variation in pressure is almost impossible. All pumps used for feeding filter presses should be provided with air vessels both on the suction and the delivery side, a pressure gauge in the top of the delivery air vessel, a safety valve adjustable to blow off at a definite pressure, and a stone trap to prevent foreign substances reaching the suction valves.

To use the same pump for both washing and feeding is not good practice, and separate pumps for these duties will be found the most economical in the long-run.

FIG. 63.—FILTER PRESS: HYDRAULIC RAM CLOSING.

As a rule filter presses are square in section, although many having circular plates are on the market. The latter have an element of additional strength which is discounted by the action of the safety valve on the delivery side. Although the tendency to buckle is less than in the case of square plates, and tight jointing is easily obtainable, yet, as the duty of a filter press is to filter, the great waste of filter cloth—nearly 25 per cent. of the cloth needed for the same diameter square plate— and the equivalent loss of filtering surface more than balance any advantage the round type may have over the square type.

Provided presses are erected level, so as to avoid any liability to leak, no special skill is required in their assembly, and after a few runs any average hand becomes quite competent to look after the plant in a satisfactory manner.

Centrifugal Machines.—The work of this machine differs from that of the filter press in that it mainly consists of removing moisture which adheres to solids, and not, as in the latter case, separating solids from an excess of moisture.

It is included amongst filtering machines because actual separation is effected by means of a filtering medium, the action of centrifugal force being merely to accelerate the operation. It finds its greatest use in the drying of crystals by throwing off the adhering mother liquor, and for this purpose is largely employed in the sugar industry.

It is capable of very extended application, but as it requires very careful workmanship, considerable skill, and experience in running it if accidents are to be avoided, the result is that many manufacturers are unable to take full advantage of this means of separation.

The centrifugal machine consists essentially of a cylinder with an adjustable perforated circumference, fixed to a vertical shaft which is rotated at a high speed.

As the cylinder and its contents rotate, the latter is driven by centrifugal force to the circumference, where the perforated screens retain the solids and pass the liquids.

Fig. 64 gives a section through a " Weston " type of centrifugal basket and outer case, showing also the central discharge with valve and inside and outside steaming arrangement.

This machine is largely made by Pott, Cassels and Williamson, Motherwell, Scotland, and Fig. 65 shows the various types of linings which this firm supply for use with the baskets. In actual practice three linings are used: a 4-mesh plain woven iron lining next the basket shell, then a 7-mesh plain woven brass lining, and an inner lining of perforated copper sheet with conical oblong holes.

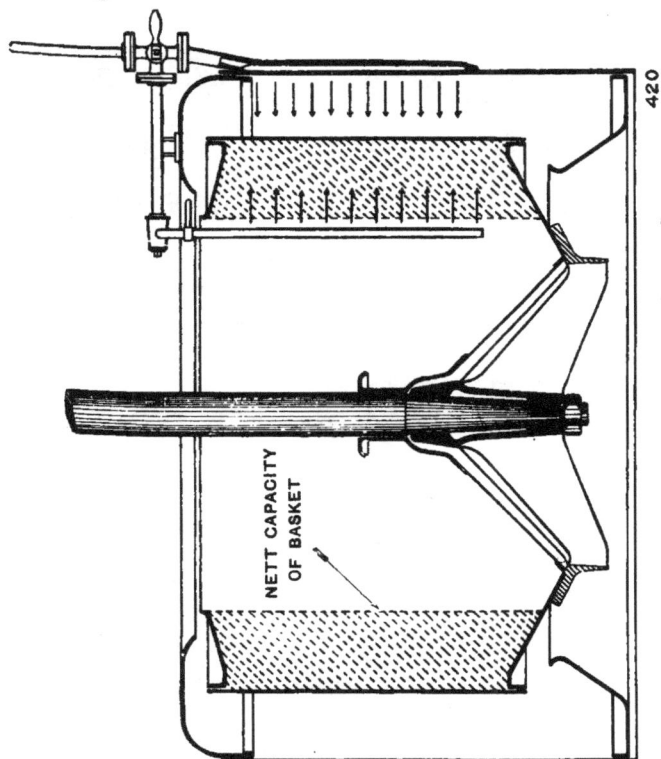

FIG. 64.—"WESTON" CENTRIFUGAL BASKET: SECTION.

For special work, linings of perforated copper sheet with conical round holes, of 26 or 30-mesh twilled woven copper or of spiral woven brass (Lieberman lining), may be used.

The spindle bearing is perhaps the most important part of the centrifugal, as it has to do continuous heavy work at high speeds with a minimum of attention. In some cases the load is about 1 ton with a speed of 750 revolutions a minute, and the machine has to work day

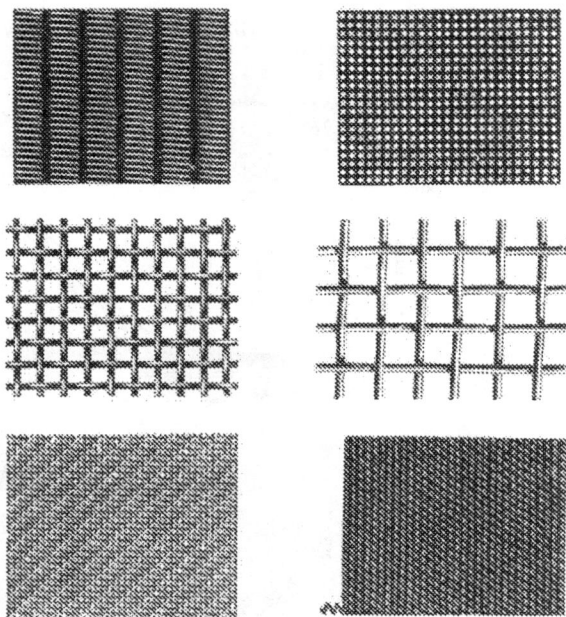

FIG. 65.—TYPES OF LININGS FOR BASKETS.

and night for months on end. It is essential, therefore, that the wearing parts should last for a very considerable time and that the cost of repairs should be small. Figs. 66 and 67 show a solid spindle with compound ball-bearing and a patent sleeve and ball-bearing spindle as made by this same firm. In this latter case both sleeve and ball-bearing run in an oil bath, and no adjustment is necessary.

The actual power used in driving a centrifugal machine is not only determined by the amount required when running at speed—a comparatively small amount in the

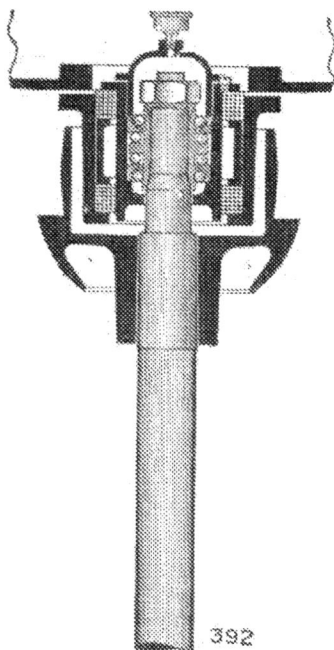

FIG. 66.—BEARING FOR CENTRIFUGAL SPINDLE.

case of a well-balanced machine—but also by the time allowed for acceleration. In practice the average time allowed for acceleration is two minutes, and the b.h.p. required is calculated on this basis. If a shorter period of acceleration is desired, there must be a corresponding increase of power available, and for a longer period a smaller amount will suffice.

Power is usually transmitted to the centrifugal from the prime mover, in most cases a steam engine, by either a belt, electric motor, or water.

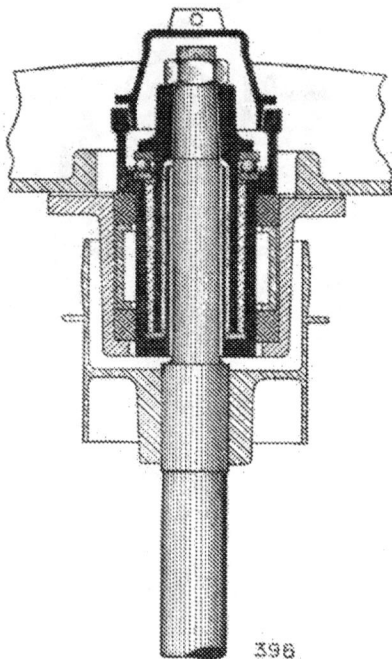

FIG. 67.—BEARING FOR CENTRIFUGAL SPINDLE.

With the belt drive the centrifugal counter-shaft is driven by an engine or, as a variation, it may be driven by an electric motor. A method which has been adopted for many years for starting up centrifugals and other high-speed machines is to transmit the power to the

FIG. 68.— CENTRIFUGAL FRICTION PULLEY

centrifugal through a friction pulley such as is shown in Fig. 68.

By this means the machine is started without undue strain on the driving belt, and may, within limits, be adjusted to give different times for getting up speed. Anyone who has the driving of a centrifugal in hand must be thoroughly acquainted with the working and adjustment of a friction pulley. The type of friction pulley illustrated can be adjusted in the following manner: First put the clutch in the off position close up to the driving arm, screw the nuts behind the springs toward the centre of the pulley as far as they will go, and remove the loose caps from the ends of the driving arms. Screw the arms into the leather-faced friction pieces until, when each friction piece is pulled as far as it will go toward the rim of the pulley there is a clearance of about $\frac{1}{16}$ inch between the leather face and the inside of the rim. This ensures that the frictions will come quite clear out of gear when the clutch is in the off position. Replace the loose caps, which are provided with toes between which flats on the screwed arms slide and prevent the rods turning. As adjusted above and without bringing the small spiral springs on the arms into play, the friction may grip too fiercely and throw off the belt or accelerate the centrifugal too quickly. In such a case screw the nuts on the rods towards the rim of the pulley so that the springs take some of the centrifugal force off the arms, thus reducing the friction and giving a lessened pull on the belt. The amount of compression on each spring should be approximately the same, and the travel of the clutch should not exceed 1 inch, which is determined by the correct setting of the loose collar.

In the case of electric-driven centrifugals the friction pulley is used with a D.C. motor, but in the case of an A.C. motor a flexible coupling is used to connect the motor to the centrifugal spindle. This type of motor has other advantages in that there are no commutators or brushes to wear and the speed is constant. It is well to note that in A.C. motors the available range of speeds depends on the frequency, and it is necessary to bear this in mind when deciding on the frequency. The table

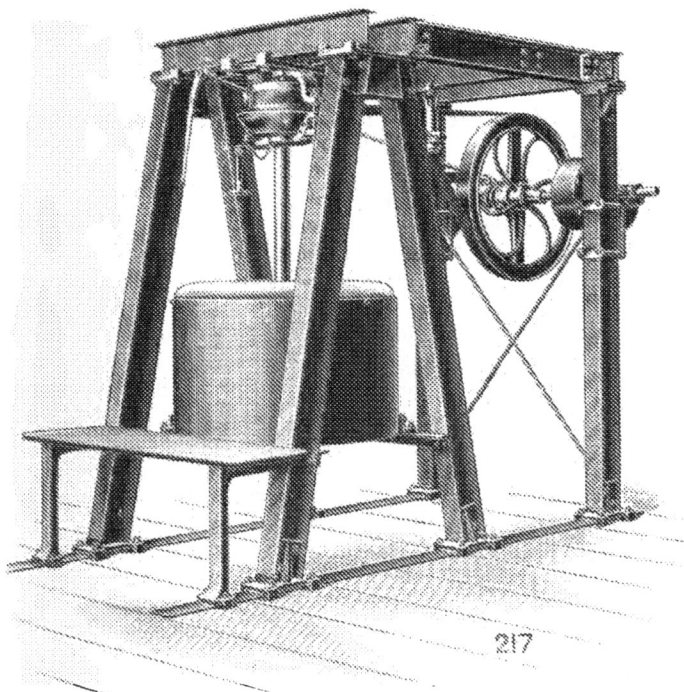

FIG. 69.—" WESTON " CENTRIFUGAL MACHINE.

given below gives the possible synchronous speeds for some common frequencies, the actual running speeds being about 4 per cent. less than the synchronous speeds.

Frequency.	Synchronous Speeds.	Diameter of Baskets of Suitable Centrifugals.
25	1,500, 750	24, 48 inches.
40	1,200, 800	30, 48, 30, 48 inches.
50	1,500, 1,000, 750	24, 36, 40, 42, 48 inches.
60	1,200, 900	30, 40, 42 inches.

Fig. 69 shows a belt-driven centrifugal having a 42-inch diameter basket, together with the necessary steel framing, etc.

In the case of water-driven centrifugals, jets of water under pressure are made to impinge on the cups of a Pelton wheel which is coupled to the centrifugal spindle. The pressure of the water is raised either by a direct-acting steam-driven duplex pump, a turbine pump driven by an engine or electric motor, or a high-duty flywheel pumping engine. The duplex pump is the cheapest and the one most commonly used; the turbine has the advantage of simplicity, constant running, and the ability to give out power approximately proportional to the power consumed; the pumping engine is most economical in steam consumption, but is the most expensive in first cost.

Fig. 70 shows a sectional drawing of a water-driven centrifugal made by Pott, Cassels and Williamson, and the following details relative thereto will serve to illustrate the various points of centrifugal machines in general.

The motor case 3-4, fitted with a cover 1, into the centre of which is secured a hollow axle 9, which does not revolve, rests on the beams 40, which form part of the framing. On the lower part of the axle a ball-bearing 10 is placed; the inner part of the ball-bearing is held firmly to the hollow axle by the nut 12, and the outer part is held in the eye of the water wheel 2 by the nut 11,

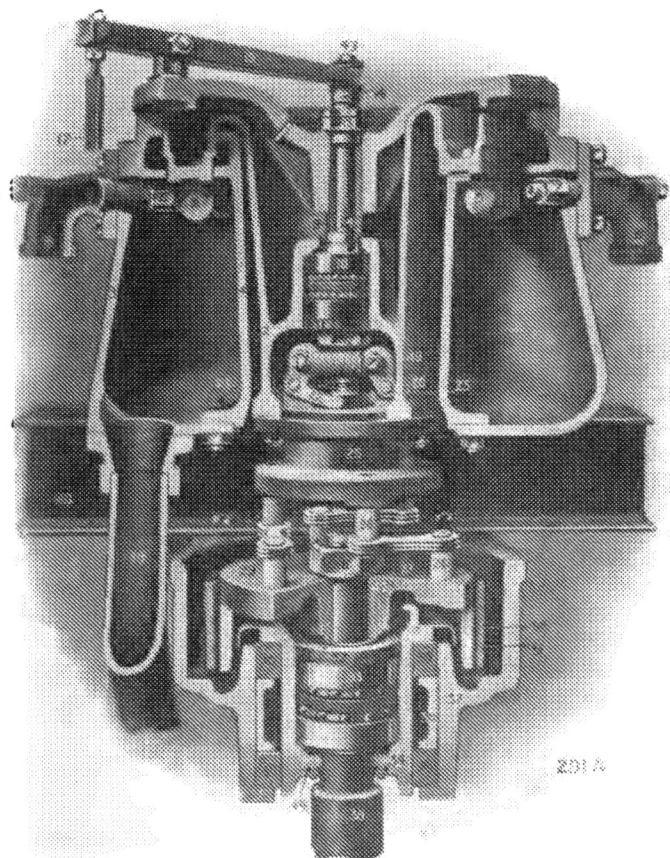

FIG. 70.—" WESTON " CENTRIFUGAL MACHINE: WATER-DRIVEN

and so revolves with it.   The upper parts of the motor case have flanges projecting towards each other, forming diaphragms to prevent the water spray from getting over the top, so that there is no possibility of the spent water going anywhere except through the return pipe 39, back to the water tank which supplies the pump for driving the machine.   The top of the water wheel, on the face of which the water cups 5 are secured, revolves between the diaphragms on the top of the motor case. To prevent any alteration in the position of the cups, they are fitted into a groove on the face of the wheel.   To obtain maximum efficiency they are carefully machined, have knife edges, and are of parabolic form, properly relieved on the bottom for the escape of the spent water. Further, as the wheel does not oscillate, these cups always maintain the same position relative to the water jets 6 and 7, which are screwed into the inlets 8, provided with inspection plugs 41.   The inlet bends 8 can be easily and quickly removed and a choked jet readily cleaned.

On the bottom of the water wheel is bolted a driver 25, which encloses the governor balls 19 in an oil-tight cavity below the ball-bearing.   This cavity is partly filled with oil through the oil cup 43, which lubricates the governor pins 23 and the ball-bearings 21 and 10.   The governor spindle 24, which is a tube for the passage of the oil, has a ball-bearing 21 at the bottom and a collar 14 at the top. The governor balls are held in the " off " position by the springs 20, which are of such strength that when the machine attains full speed the centrifugal force causes the balls to fly outwards and move up the spindle 24 by means of the levers 22.   On the top of the motor case cover is a fulcrum 16 for the lever 15, at the shorter end of which is a swivelling cross-head 18, through which passes the governor rod 17, adjusted and secured by two nuts.

When the machine attains full speed the governor rod is forced by means of the levers, and releases a trigger which cuts off the water from the accelerating jet 6, leaving the smaller jet 7 to maintain operations.

To the underside of the beam is attached the centrifugal suspending block 37, into which are fitted india-rubber buffer rings 35, separated by a loose cast-iron ring 36. Thus top and bottom buffers support the weight of the basket, which is attached to the lower end of the centrifugal spindle 38. By this means great resiliency and steadiness are obtained when the machine is running with either a balanced or an unbalanced load, and also, as the buffers are separated by a loose ring, any wear on the bottom buffer is compensated by the compression caused and is self-adjusting.

The ball-bearing housing 34 fits inside the india-rubber buffers and contains the compound ball-bearing 33, secured by a nut 32; the inner part of the ball-bearing is secured to the centrifugal spindle by the top nut 28 through the brake casting 29. To permit of the oscillation of the centrifugal spindle and the basket, the water wheel, which does not oscillate, is connected to the brake pulley on the top of the spindle by leather links 27, the eyes of which are slipped over the points of the driving pins 26. Thus a strong flexible coupling is formed and one which permits of the motor or centrifugal being detached as desired by simply slipping off the links. The brake band is supported by angle iron feet which rest on a flange in the bracket 37, so there is no possibility of the band drooping unequally. The feet on the brake band are so arranged that when the brake is off an equal space is left all round between the band and the pulley.

As is well known, the power to accelerate a machine to full speed quickly is much greater than is required to maintain it at full speed; consequently each centrifugal is provided with two water jets, a large one and a small one. When starting the machine both jets are required, and when full speed is reached the small jet only is

necessary to maintain full speed.   It sometimes happens
that when the centrifugals are worked irregularly all
the machines may be accelerating at the same time,
requiring an amount of water largely in excess of normal
requirements, and for this reason pumps have hitherto
been made very large.

Fig. 71 shows two machines interlocked by a special
gear made by this firm, and so arranged that not more
than one half of the machines can be accelerated at the
same time, thereby reducing the size of the pump considerably, without in any way reducing the output.   In

FIG. 71.—INTERLOCKING GEAR FOR WATER-DRIVEN CENTRIFUGAL:
SECTION.

ordinary practice each pair of machines is interlocked,
so that when one machine is started the other machine
cannot be started until the first has attained full speed.
As the machines are usually arranged to accelerate
in two minutes, and the cycle of operations will occupy
at least six minutes, the interlocking gear ensures the
machines being worked in proper rotation without
interfering with the output.   The advantages of this
method will be evident from the following comparison
of the maximum pump demand for a set of machines with
and without this interlocking gear.

For instance, a set of machines which are interlocked in pairs, and require a pump 9¾ inches diameter, would require a pump 12¼ inches diameter otherwise, provided the maximum pump speed is the same in both cases. In other words, the maximum pump demand is 55 per cent. more. When a more rapid acceleration is required, three machines are interlocked, in which case a pump 11¼ inches diameter would do the work instead of a pump 17 inches diameter, which represents an increase of 130 per cent. It should also be remembered that in most cases the steam cylinder is at least twice the diameter of the pump, and the smaller the pump the smaller the steam cylinder.

The following cases extracted from a list given by this firm affords useful comparisons.

Diameter and Depth of Basket (*Inches*).	R.P.M.	Average B.H.P.	Capacity (*Cubic Feet*).
24 × 14	1,500	2	1·85
48 × 20	750	5·25	10·7
60 × 24	600	10	16·6

CHAPTER IV

DRYERS AND EVAPORATORS.

ALTHOUGH the bulk of the moisture in a material has been removed by means of the filter press or the centrifugal machine, a certain amount still adheres, which can only be removed by evaporation in contact with air heated to as high a temperature as is possible, consistent with economy and the nature of the material.

The three systems of drying most commonly in use are:

1. By direct heat from a fire.
2. By radiated heat from steam pipes.
3. By warm air circulation.

Machinery for using direct heat from a fire has only a limited application, owing to fire risk and the liability of damage to the material, although for substances such as sand the method is very effective. Wherever possible waste heat should be utilized, so that in the case of materials which are not easily burned or scorched or damaged by contact with gases the flue heater provides an effective and economical drying means.

A common form of flue heater consists of a cast steel or iron trough placed over a flue or furnace, the material being propelled from one end of the trough to the other by a worm, which is also made to act as a stirrer or turner over. These troughs are of varying diameter and lengths to suit the material to be treated, and can be open or enclosed and connected with a fan if necessary.

The lack of any very effective control is one of the great disadvantages of this method, so that it is not to be

98

wondered at that the ease of control of steam has rendered that substance the principal heating agent in dryers and evaporators, apart from the fact that it allows of the utilization of much waste heat.

Steam may be used for drying or evaporating by being passed through pipes immersed in the material to be dried, by forming a steam jacket round the container, by heating an inner drum round which the material passes, or by a combination of these methods. Frequently the drying process is combined with mixing and milling, according to the nature of the material used.

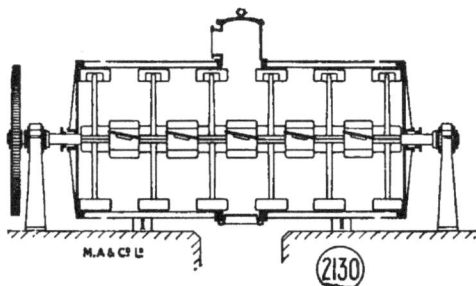

FIG. 72.—" FIRMAN " DRYER: LONGITUDINAL SECTION.

The rotary form of dryer is undoubtedly one which finds the greatest application in the chemical industry. In its simplest form it consists of a cylinder or drum, steam-heated, containing the material, which can be rapidly and uniformly presented to the steam heat by rotating some part of the apparatus. Materials such as slaughter-house refuse, blood, offal, condemned meat, fish and vegetable matter can be turned into valuable manure by means of a dryer as shown in Figs. 72 and 73.

This is a " Firman " type made by Manlove, Alliott and Co., Ltd., Nottingham, which is designed for taking semi-liquid material and delivering it mixed and dried ready for the market. It consists of a horizontal steam-jacketed cylinder, the internal circumference of which is continually swept by moving scrapers or paddles. The material being treated is kept in constant motion, being lifted up, turned over, and allowed to fall again, presenting fresh portions to the heated surface and giving the steam and vapour a better opportunity of escaping.

The temperature at which the material is dried can be controlled by regulating the steam pressure, and so no injury is caused by excessive temperature, as is frequently

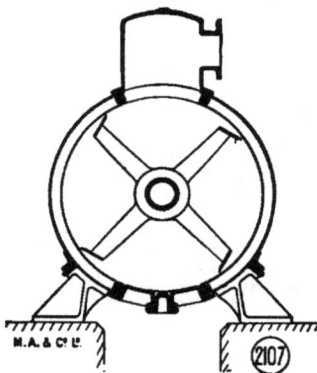

FIG. 78.—" FIRMAN " DRYER: CROSS SECTION.

the case with fire-heated machines. The body of the machine consists of a double cylindrical shell suitably stayed to withstand a working pressure of 40 pounds per square inch, the space between the two shells forming a steam jacket. A massive cast-iron plate at each end of the drum is fitted with a gland, bracket, and plummer block for supporting the central shaft of mild steel, which carries cast-iron arms having knife-edge steel scrapers at their outer ends. A charging door is fixed in the top or one end of the machine, an outlet door in the

bottom, and a vapour outlet at the top.   A draining valve
connected in two places and a steam inlet valve, together
with the usual pressure gauges and safety valve, complete
the equipment of the steam jacket.

Fig. 74 shows a cross-section of a dryer fitted with a
central steam drum extending the greater part of its
length, and Fig. 75 that of a combination of the steam
jacket and internal drum.   These machines are suitable
for granular substances such as beer and distillery grains,
earthy and other colours, coal, sawdust, peat, etc., or
substances which have a low moisture content.

FIG. 74. — " HERSEY " ROTARY   FIG. 75.—COMBINATION ROTARY
      DRYER: CROSS SECTION.   DRYER: CROSS SECTION.

In these cases the plan of operation is for a current of
air heated by passing through a steam-heated air heater
to be drawn through the cylinder by means of a fan.
The material is fed by hand or automatically, and is
lifted by the shelves and rained down through the hot
air, to which it gives up its moisture.   Owing to the
cylinder being set at an angle, the material passes down
in the opposite direction to the air current, so that the
driest material comes in contact with fresh hot air,
effectively removing the last traces of moisture, while
before the air escapes it comes in contact with the fresh

wet product, which cools it to the lowest temperature
consistent with the proper carrying away of the absorbed
moisture.

The temperature of the air used in this class of plant
is about 120° to 200° F.—the temperature in general

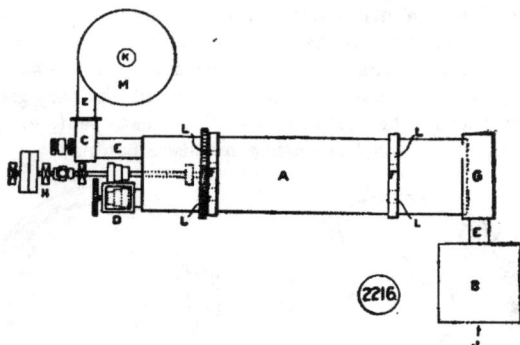

FIG. 76.—G.A. DRYING PLANT: PARALLEL DRIVE.

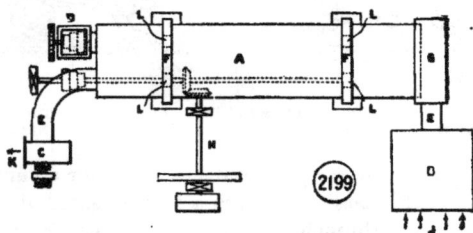

FIG. 77.—G.A. DRYING PLANT: RIGHT ANGLE DRIVE.

A, rotary dryer; B, air heater; C, fan; D, feed apparatus;
E, air pipes; F, path rings; G, discharge hood; H, driving
pulleys and gear; J, air inlet; K, air outlet; L, rollers;
M, dust separator.

use being about 150° to 180° F. measured close to the
heater.   In dealing with material of a powdery or dusty
nature the discharge from the fan may be led to a large

settling room, or a cyclone or other separator may be used.

The main factors governing the size of plant for a given output are—(1) The quantity to be dealt with per hour; (2) the initial and final moisture percentages; (3) the speed of the air current which can be employed without carrying away too much material in the form of dust; (4) the temperature which can be safely employed; and (5) the ease with which the material gives up its moisture. Where a very low final moisture is desired, or the material dries slowly, it is better to have an extra long dryer.

Figs. 76 and 77 show the general arrangement of drying plants with parallel and right-angle driving gear respectively.

Drying by warm-air circulation forms the basis of the Sturtevant system. A fan passes a volume of air through a self-contained heater placed outside a drying room; this air is led through a system of pipes and distributed in the drying room, a positive circulation of warm air being maintained, and the heat necessary for evaporating the moisture in the materials is carried into every part of the room. The advantages claimed for this system are—(1) The temperature can be varied without affecting the volume of the air circulated; (2) the humidity of the air supply can be varied by recirculating part of the moist air from the drying room; (3) the volume of air circulated can be regulated by varying the speed of the fan or by dampers.

This method is particularly suitable for the drying of wool, flocks, rags, fibre, and similar substances, where it is important that there should be no risk of fire.

When air is brought in contact with a wet substance some portion of the moisture is absorbed by the air, which has an increased capacity with increased temperature. For any particular temperature there is a limit to the amount of moisture the air will absorb, and when this limit is reached the air is said to be saturated. Saturated air is obviously useless as a drying agent, but by raising the temperature of the saturated air it becomes capable of taking up more moisture before it again becomes saturated. In other words, the higher the temperature of the air, the better drying agent it becomes. The limit of temperature usable depends, of course, upon the nature of the material to be dried.

FIG. 78.—RECORDING HYGROMETER.

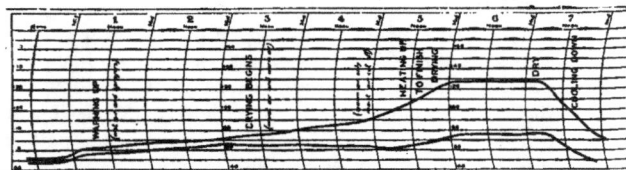

FIG. 79.—TYPICAL GUIDE CHART.

An important part of the Sturtevant system is the recording of the humidity of the air in the drying rooms. This is performed by a self-recording hygrometer which has wet and dry bulbs (metallic expanding coils) which actuate two pens against a revolving drum carrying a guide chart, as shown in Figs. 78 and 79.

The following table is then found necessary in order to calculate the humidity of the air used:

DIFFERENCE BETWEEN WET BULB AND DRY BULB IN
DEGREES FAHR.

Dry Bulb ° F.	2	4	6	8	10	12	14	16	18	20
	Percentage of Moisture in Air.									
60	89	78	68	58	49	40	31	22	14	6
70	90	81	72	64	56	48	40	33	26	20
80	91	83	76	68	61	51	47	41	35	29
90	92	85	78	71	65	59	53	47	42	37
100	93	86	80	74	68	62	57	52	47	42
110	94	87	81	76	70	65	60	55	50	46
120	94	88	82	77	72	67	62	58	54	49
130	94	89	84	78	74	69	65	60	56	52
140	95	89	84	80	75	71	66	62	58	55
150	95	90	85	80	76	72	68	64	60	56
160	95	90	86	81	77	73	69	66	62	58
170	96	91	87	82	78	74	70	67	63	60

The actual amount of water in a cubic foot of air can then be found from a chart, as shown in Fig. 80.

For example:

The dry bulb thermometer registers 140° F.
The wet bulb thermometer registers 122° F.
Difference......................18° F.
Percentage of moisture (from table), 58 per cent.
Moisture per cubic foot (from chart), 33½ grains.

The process of drying timber provides an example of the method of working and flexibility of this system. The seasoning of timber not only affects the moisture content, but also goes deeply into the quality of the wood —its workability, its cell strength, etc.—for in green wood the moisture is divided between the cells and the cell walls. The free water in the cells or pores can be removed without affecting the physical structure of the wood, but it is otherwise in the case of the water in the cell walls. Successful drying of timber depends upon the following fundamental conditions: Gentle heat in combination with sufficient moisture to prevent case hardening; an ample circulation of air, so that the humid heat is carried to every part of the timber. The moisture should be evaporated just as quickly as it comes to the surface of the timber, and there should be no great temperature drop throughout the pile or in each piece of timber. By this system, when one particular kind of wood has been dried, the particular guide chart—for the same thicknesses of wood—can be used subsequently.

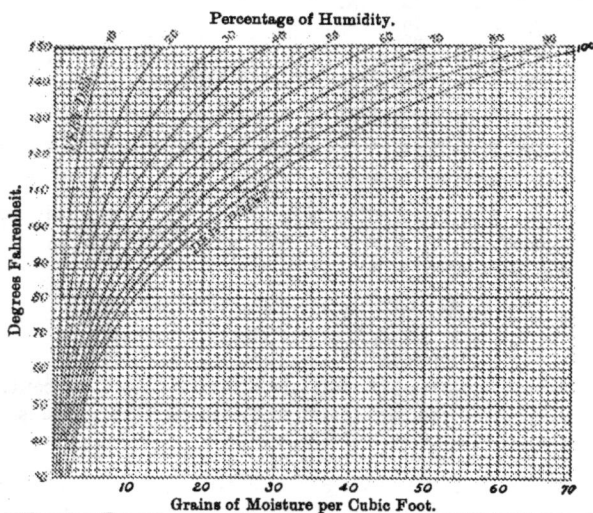

FIG. 80.—HYGROMETRIC CHART.

All that need be done is to put the chart on the drum of the hygrometer and manipulate the dampers so that the pens of the recorder will follow the lines of the chart. Warm and cold air dampers govern the top line of the diagram, and the moist air damper and sprayers control the bottom line.

For large and regular outputs of planks, barrel staves, and suchlike, a progressive type of dryer is used. This consists of one or more tunnels of brick or concrete to reduce heat losses by air leakage to a minimum, with rails running longitudinally, the wood being piled on trucks which move through the tunnel against the air current. The arrangement is such that whilst the wood that is nearly dry is in contact with warm dry air from the fan, the green wood at the other end meets moist tepid air which has been cooled and moistened by passing over the wood of the first waggons. Whenever a load is taken out dry, a load of new wood is put in at the other end, all the waggons moving one stage forward; thus, progressively, the wood is dried in a suitable atmosphere.

The air ducts are usually placed underground and constructed with brick sides and concrete base, the top being arched over with brick or covered with stone or concrete.

The drying apparatus consists of—(1) A fan for producing circulation driven by a direct coupled steam engine, electric motor, or belt; (2) a heater for heating the air by live or exhaust steam or a combination of both; (3) a steam trap to prevent the passage of steam, or an automatically controlled pump and receiver to return the water of condensation to the boiler; (4) steam and water sprays for supplying such additional moisture to the air as is necessary.

For drying different thicknesses and sizes of timber and also different kinds of wood, especially hard woods, three main distributing ducts are provided in the compartment dryer for the supply of warm dry air, cold dry air, and moist air. The supply to each compartment can be controlled from a central board placed conveniently outside, as shown in Fig. 81, which is a plan and elevation of the triple drying system.

The timber, if a hard wood, is first subjected to the action of moist cool air, and the humidity is maintained while the temperature is slowly raised, thus gradually removing water from the heart of the timber without drying the skin. The humidity is then gradually decreased and the temperature raised, and when the drying is nearly complete warm dry air is admitted to finish off.

In the above illustrated system the fan takes fresh air, part of which is delivered through a by-pass valve,

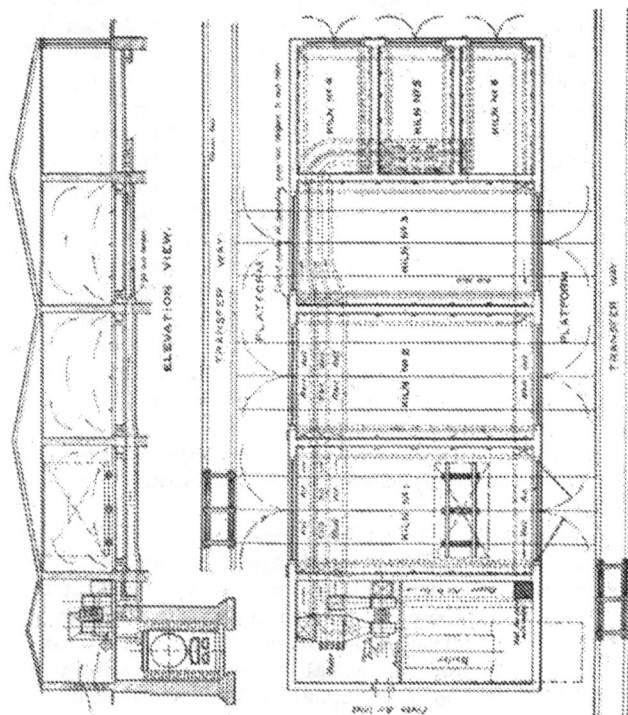

FIG. 81.—STURTEVANT TRIPLE DUCT DRYER: SECTION.

into the cold air duct, the remainder passing through a heater to the hot air duct. A second fan draws the moist air from the drying chambers through the return duct, and delivers any desired percentage of it to the moist air supply duct.

By adapting the apparatus to suit the different materials this system has been successfully used to dry asbestos, casein, copra, explosives, fruit, glue, leather, phosphates, rubber, soap, sugar, white lead, and wood pulp, to mention but a few.

Vacuum Dryers.—A vacuum dryer is a machine in which material is dried under reduced pressure, whereby a considerable reduction in time and expense is effected. The temperature at which a liquid changes rapidly and violently into a vapour, or, in other words, boils, depends upon the pressure upon its surface. By reducing the pressure in a drying machine and maintaining it at any desired level, the moisture, usually water, in the material may be briskly evaporated at a convenient low temperature, so that waste steam of low temperature may be usefully employed. Not only is this method quick and economical, but it possesses the further advantage that it is possible to dry rapidly substances which would be decomposed or otherwise injured if raised above a certain definite temperature.

There are three main types of vacuum dryers adapted for different classes of materials, as follows: (1) The shelf type for materials which do not require stirring; (2) the rotary type for materials which require stirring; and (3) the drum type for material which readily forms a film on the drying surface.

Fig. 82 is an illustration of a vacuum shelf dryer, made by Francis Shaw and Co., Ltd., Manchester. The chamber itself is a heavily ribbed cast-iron box, in one or more sections as necessary, having the faces where the joints are to be made, accurately machined. The doors are also of cast iron, securely attached to the body by double swing hinges and accurately balanced for easy manipulation. To obtain an air-tight joint when the door is closed, a rubber ring is fitted into a groove on the inside skirting and makes contact with the machined facing on the body. For the purpose of control, inspection windows are fitted so that the material can be kept under observation during the drying operation.

FIG. 82.—VACUUM SHELF DRYER

The heating chests or shelves are made of rolled mild steel plates, flush-riveted at the edges through a welted ring, and stayed all over to withstand a working pressure of 60 pounds per square inch. Each shelf has its own independent steam feed and exhaust connections, made of stout hydraulic piping bent to allow for expansion and contraction and coupled to mains fitted in recesses. These shelves are arranged to give complete drainage and allow of uniform heating. A cast-iron vapour pipe leads from the top of the stove to the head of the condenser, which is of the vertical multitubular type, containing copper, brass, or iron tubes expanded into end plates at top and bottom. The condenser case is of cast iron fitted with condensing water inlet at the bottom and outlet at the top, together with all necessary valve fittings. The receiver is a cast-iron vessel divided into two compartments connected by a by-pass valve, by which the liquid in the lower compartment can be drawn off without breaking the vacuum throughout. Inspection windows are also fitted to this vessel so that the drops of water falling from the condenser can be seen during the drying operation.

The method of working a shelf dryer first of all is to warm up the shelves by admitting steam or hot water, as the case may be. The material to be dried is placed to a depth of 1 to $1\frac{1}{2}$ inches on trays of galvanized or enamelled iron, wire netting, copper, aluminium, or earthenware, having a depth not greater than 2 inches, as the shelves are about $2\frac{1}{2}$ inches apart. The trays are then placed on the shelves and the doors closed by means of handwheels. Provided that the doors fit perfectly, the handwheels can be swung clear as soon as the gauge registers 10 inches, when the excess external pressure is sufficient to hold the doors in place. The vacuum pump is now started, and in a short time drops of water can be seen through the inspection window falling into the bottom chamber of the receiver. Cold water is then admitted to the condenser, and a sufficient flow maintained to keep the bottom of the condenser cool to the touch. To obtain a constant drying temperature inside the stove the steam or hot water inlet must be adjusted as required. The end point of the drying operation is indicated by the fact that no more drops of water are observed falling into the receiver, or by a decided rise of temperature registered by a thermometer in the body of the stove. The operation is then stopped by closing the vacuum valve and shutting down the pump. To avoid loss of material the air must be admitted gradually through a vacuum break valve in the door, until zero is recorded on the vacuum gauge, when the material can be taken out and the stove recharged with a fresh set of shelves already prepared.

These stoves are made in all sizes from 10 square feet up to 3,000 square feet of heating surface, and may be combined in the form of a battery if required.

This type of machine is used for drying aniline dyes, pigment colours, explosives, fine chemicals, malt extract, carbon brushes, white lead, beta-naphthol, salicylic acid, drugs, mica sheets, foodstuffs, etc.

For substances which have a tendency to form a film on the surface of the dryer, and for such substances as sulphate of zinc, nitrate of ammonia, dyewood and tannin extracts, glue solutions, albuminous substances, milk, yeast, pastes, eggs, vegetable and meat extracts, etc., the drum dryer is the more suitable machine.

Fig. 83 illustrates a vacuum drum dryer as made by J. P. Devine and Co., Buffalo, working under what is known as the Passburg system. Briefly, the apparatus consists of a cast-iron outer casing, in some cases tinned inside or lined with copper or other suitable material, inside which revolves a hollow drum or drums made of cast iron, gun-metal, or bronze. This drum is carefully balanced, heated internally by hot water or steam, and the outside is machined and polished. If hot water is used as the heating medium, drying can be conducted at as low a temperature as 63° F., and in the case of steam 1·2 pounds of steam—including steam used for motive power if the exhaust is used for heating the chamber—will evaporate about 1 pound of water. The inlet supply is regulated so that the material inside the casing remains at a constant level, and shield rings prevent the material from coming in contact with the end of the drum.

A rapid and uniform drying is effected, because the wet material is spread upon the steam-heated polished

FIG. 83.—VACUUM DRUM DRYER.

drum in a thin film of $\frac{1}{120}$ inch or less. The conduction of heat through the metal drum to the material on its surface is very rapid, and the water is changed under vacuum into vapour of about 104° F. to 122° F. Materials containing as much as 88 per cent. of water are dried in eight seconds without overheating, and the water is evaporated from the material at a temperature of from

8

117° F. to 96° F., according to the vacuum in the apparatus of 26¾ inches to 28¼ inches.

It is well known that the greater the difference in temperature between two materials placed in contact, the more rapid is the flow of heat from the hotter substance to the cooler, so that if the water in the material is kept at a low boiling point by maintaining a high vacuum, the heat from the steam is transmitted more rapidly than if the boiling point be several degrees higher. Even though the drum be heated by steam of 230° F. or over, the material being dried cannot get higher than the temperature at which the water boils

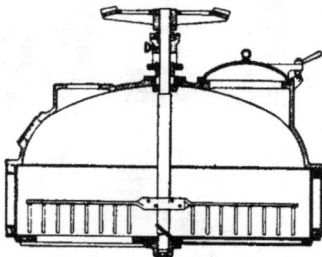

FIG. 84.—VACUUM " JOHNSTONE " DRYER: SECTION.

under vacuum, because the heat is used to convert the water into steam, and thereby rapidly evaporates it out of the material.

The rotary vacuum dryer follows the lines of the non-vacuum type previously described, but with the addition of an evacuating plant. Here, again, the machine usually combines the work of a ball mill or mixer, or both, with that of drying.

Fig. 84 shows a section of the " Johnstone " dryer made by Manlove, Alliott and Co., Nottingham. It consists of an enclosed vacuum-tight vessel with a dome-shaped cover which carries a scraper, agitator, and driving gear. In the cover is placed a large circular charging door, which is provided with shackles for tightening on to a joint ring of asbestos or rubber. The body of the machine has steam-jacketed parallel sides and flat bottom capable of withstanding 40 to 60 pounds per square inch. In the bottom of the machine is a rectangular hinged and balanced door, opening downwards, for discharging the dried material as required. A thorough and speedy drying is ensured, as the material is continually broken up and turned over by revolving scrapers and rakes, thus preventing an impervious crust forming and hindering evaporation.

Fig. 85 shows a combined vacuum dryer, mixer and ball mill, made by Francis Shaw and Co., for treating delicate chemicals, organic acids, and all substances where metallic contamination is to be avoided. It consists of a fixed steel casing designed for heating by steam or gas, inside which is fitted a one-piece stoneware vessel, supported in a steel cage having a hollow trunnion at one end connected to the vacuum pump.

One of the disadvantages of many drying machines is the loss of time in charging and discharging, and several makers have evolved a more or less efficient continuous apparatus.

Fig. 86 shows a continuous "cone" vacuum drying plant made by this same firm. The body of the machine consists of a steel cylinder, jacketed all round for steam heating, and having a hopper fitted at the top of one end. At the base of the hopper is a vacuum-tight rotary device for receiving the wet material and delivering it to the conveyor worm inside the dryer. A similar hopper is provided on the underside of the machine, connected through a wheel valve to a receiver provided with inspection windows and manhole door. The cone is made of light sheet steel mounted on a hollow shaft extending through stuffing box bearings in the ends of the outer steel body to the driving gears and steam supply at one

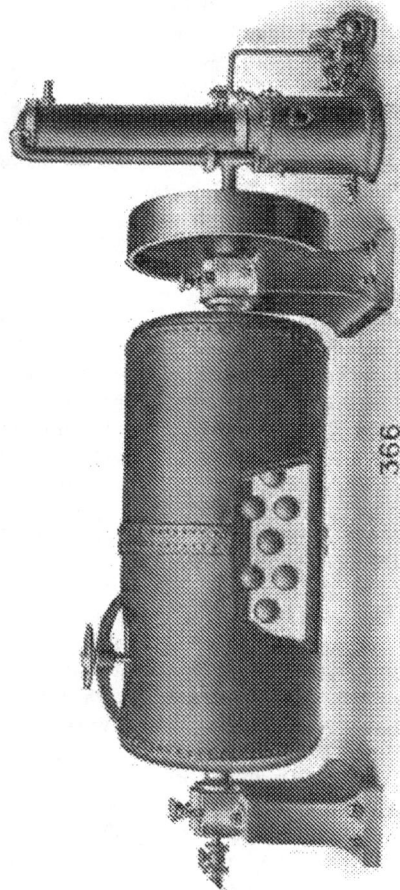

FIG. 85.—VACUUM DRYER: MIXER AND BALL MILL.

Fig. 86.—Continuous Cone Vacuum Dryer

end, and to the steam exhaust main at the other. Steel chutes are fitted to the ends of the cone inside the body for supplying the cone with the material to be dried, and for guiding the material to the receiver. Both the cone and the automatic feeding device are geared to the same shaft, to ensure a uniform feed.

In working this machine steam is admitted, and then as high a vacuum as possible is obtained throughout the apparatus with receiver valve wide open. The material is fed into the hopper, from which the worm conveyor inside obtains a uniform feed; drying commences at once, and continues as the material travels along the worm, falls down the chute to the small end of the cone and along to the wider end, where a chute delivers it to the discharge hopper in a dry condition. When the receiver is nearly full it can be discharged by cutting it off by means of the wheel valve and admitting air until the vacuum is gone.

One great advantage of vacuum dryers is that where valuable solvents have been used they can be completely recovered in the condenser. This part of the subject will be more fully dealt with at a later stage.

Evaporators are machines used for recovering solids which are dissolved in liquids by turning the latter into vapour. The methods employed may be divided into four classes, as follows: (1) Spontaneous evaporation; (2) direct heating; (3) steam heating; and (4) reduced pressure.

Spontaneous evaporation can be conducted successfully in countries which have a definite period of hot dry weather. Natural brines are frequently evaporated by pumping them into ponds having a depth of about 2 feet and a surface of several acres, at a rate equal to the rate of evaporation. In large installations as much as 5,000 gallons per minute of fresh brine is required to make up the evaporation losses, and the concentrate is led into a separate pond and harvested.

In the direct-heat method, flames or hot waste gases may be employed either to warm a vessel containing the liquid, from underneath, or by being made to pass over the surface of the liquid. Where the presence of a certain amount of impurity does not matter, the latter method can be adopted in preference to the former method, which is not so economical of the available heat. Although the open kettle is practically obsolete to-day, it is interesting to note the evolution of the process and the causes which led to its abandonment. From the simple domestic copper heated by an open fire there evolved the "block" or arrangement of as many as 100 kettles of about 150 gallons capacity, arranged in one or more rows in a flue or arches terminating in a chimney. To prevent overheating of the kettles nearest the fire, arches were built underneath them as a protection, thereby causing a certain amount of heat to be wasted. Beyond the grate the arches were built with air spaces between them, which increased in size as the distance increased, and the flues decreased in depth to about 6 inches under the kettles next the chimney. Forced draught then became necessary. Various difficulties arose which, combined with the fact that it was found that for the fuel used only about two-thirds of the amount of water was evaporated as would be effected by a proper boiler, led to the method being discarded.

Evaporation by means of direct heat is, however, still employed, but the kettles are replaced by open pans arranged as shown in Fig. 87.

The pans are of riveted wrought-iron plates $\frac{1}{4}$ inch thick, having flaring sides and divided into two or more compartments, known as the front and back pans. The pans are about 100 feet long, 25 feet wide, and 1 foot deep, and the method of working is to run the liquid into the back pan, where, after being heated by the hot gases from the fire, it is siphoned into the front pan and harvested, the solid being removed to the sides and allowed

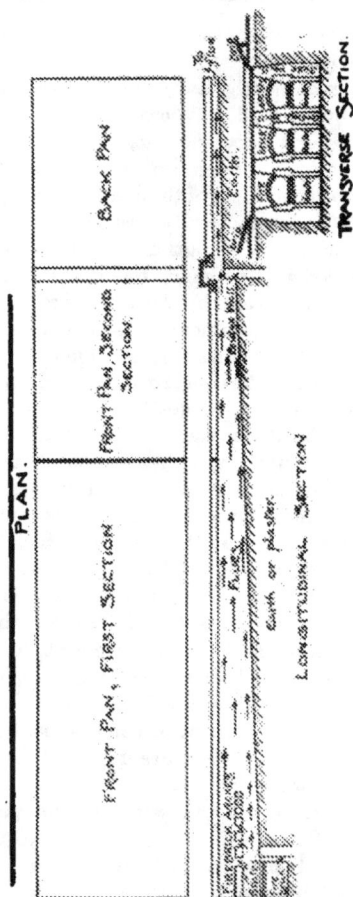

FIG. 87.—EVAPORATING PAN: OPEN TYPE.

to drain. Heating may be effected by oil, coal, or gas, according to circumstances.

Evaporation by means of steam heating is very widely

used, owing to the ease of control and the absence of any risk of damage to the product by overheating. In its simplest form a steam-heated evaporator consists of a vessel having steam pipes immersed in the liquid to be evaporated.

In the salt industry brine is evaporated in a steam-heated vessel known as a grainer—a long, narrow, shallow vat built of wood or metal supported on a framework, or of cement or concrete supported on a foundation of sand. Wooden grainers made from white pine caulked with oakum have been found to keep quite tight under the great differences of temperature encountered.

The reinforced concrete type is monolithic, having no expansion joints, and is usually provided with mechanical raking devices. The walls are 5 to 7 inches thick, the bottom 4 to 6 inches thick with $\frac{1}{4}$ inch steel-bar reinforcement, the whole resting on a sand bed which gives uniform support and reduces heat losses. An average size grainer is about 150 feet long, 12 feet wide, and 2 feet deep, having four to eight steam pipes, 3 to 5 inches in diameter, suspended about 12 inches above the bottom.

The principle on which the grainer works is as follows: If a solution of several salts is concentrated by evaporation at a given temperature, the salts will be deposited as they reach their saturation point. This will depend upon their initial concentration and solubility and the effects of the presence of other solutes. If, when the first salt is deposited and before the second salt begins to form, more of the original solution is added, the concentration of the first salt is decreased less than that of the others. Thus a thick deposit of the first salt can be formed by the continuous addition and evaporation of the original solution, until the concentration of the other salts reaches saturation point and they also begin to form. This method of separation should be carefully distinguished from the method known as fractional crystallization.

Steam-Jacketed Pans.—For general use in the chemical industry, steam-jacketed pans, such as shown in Fig. 88

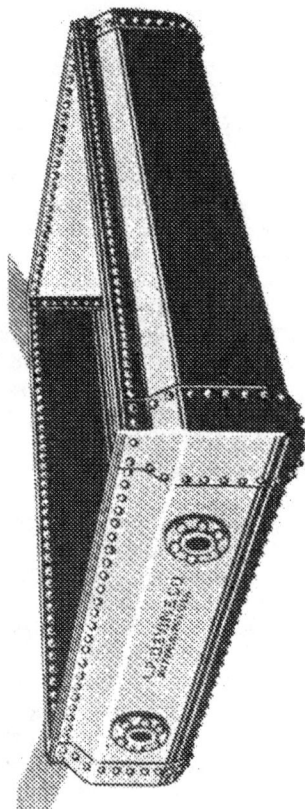

FIG. 88.—CRYSTALLIZING PAN: STEAM-HEATED.

and made by J. P. Devine and Co., are built in every size and shape, designed especially for the service under which they are to operate. They are usually made of sheet steel, but have been built of copper and cast iron, rectangular or cylindrical, welded or riveted. Proper openings are provided for steam inlets and condensed water outlets. They are usually built shallow, to allow for the greatest possible heating surface and to ensure

FIG. 89.—TILTING KETTLE.

maximum evaporation and facilitate the removal of finished material. Where high-pressure steam is used, the jacket must be properly stay-bolted.

Steam-Jacketed Kettles.—A very convenient type of apparatus, of which an example made by the above firm is shown in Fig. 89, is known as the " tilting kettle." These large kettles are usually arranged on a rigid struc-

tural steel support of the necessary height to allow for
their emptying into a truck or similar device. They may
be equipped with a cover and a stirring gear of the horse-
shoe, grate, or propeller type, driven through bevel
gears and fixed and loose pulleys, which are rigidly

FIG. 90.—ASPINALL STEAM EVAPORATING PAN.

fixed to a heavy bridge on the top flange of the kettle.
The tilting device is formed of a worm gear and worm,
the latter operated by a handwheel. The kettle proper
is supported by hollow trunnions, which also provide
the steam inlet and condensed water outlet.

The common form of kettle is made of copper or cast iron, with the jacket covering about half the kettle and with provision made for the admission of thermometers and for draining. In special cases kettles are made with enamelled linings, or of special alloys for resisting corrosive liquors or preventing contamination of the materials by the metal of the kettle.

The method of working is to open the valve to the drain pipe so that the first condensations can escape and bumping be prevented. The exhaust valve is then opened, and the steam gradually turned on at the inlet valve until a good jet of dry steam issues from the drip, which is then closed and the steam regulated to give the desired temperature for evaporation.

This type of evaporator finds very great use in the dye, paint, textile, and canning and preserving industries.

Fig. 90 shows a steam evaporating pan used in the sugar industry, made by Blair, Campbell and McLean, Ltd., Glasgow. The shell and conical top is of wrought iron, and the bottom of cast iron fixed to the top by an angle iron ring and bolts. The heating drum is of wrought iron or brass, with solid drawn brass tubes expanded into the same and beaded over, and with a large circulating tube in the centre. The drum is designed for a working pressure of 60 pounds per square inch, and is fitted with an eye-bolt for lifting and cleaning.

Fig. 91 shows a Wetzel evaporating pan made by the same firm, which is used for concentrating syrup. The heating surface is obtained by a seamless copper helical coil, the ends of which are attached to the cast-iron trunnion pipe which admits and discharges the steam, and is fitted with spur gearing driven by a pulley.

The Vacuum Pan.—The vacuum pan may be regarded as a modification of the vacuum dryer which has been considered previously. At the present time it is almost universal practice to evaporate under vacuum those substances which are liable to damage either by a high temperature or by the presence of air. Although the temperature at which a liquid boils depends upon the

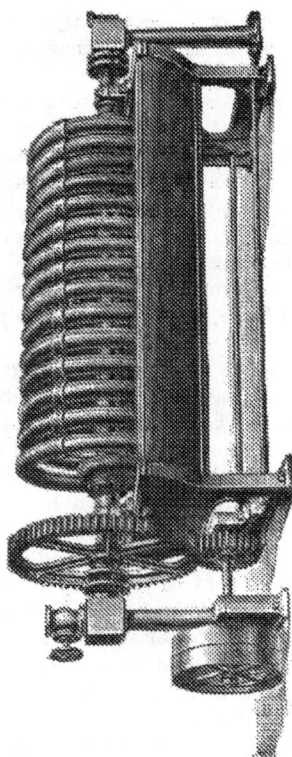

Fig. 91.—Wetzel Evaporating Pan

pressure to which it is subjected, in practice there are limits to the range of temperature and pressure available. As mere evaporation in a vacuum does not necessarily mean economy, the utilization of exhaust steam fixes the upper limit of temperature at about 225° F.; the lower limit, at about 125° F., is determined by the cost of maintaining the required vacuum. Obviously, with a given amount of heat available, better results are to be obtained with a vacuum than with normal pressure, and the greater the vacuum, the greater the amount of liquid evaporated. Against this increase in evaporated material must be reckoned the additional cost of the special plant and the cost of maintaining the vacuum, both of which must be taken into account when comparing vacuum evaporation with the open pan process.

A vacuum-pan installation consists of three parts: (1) The vacuum pan; (2) the condenser, and (3) the receiver and pump. The vacuum pans are usually vertical cylinders having conical ends, the whole being constructed of iron, steel, or copper, and provided with inlet and discharge holes, thermometers, vacuum gauge, test cocks, liquor gauge, etc. These pans are seldom less than 9 feet in diameter, and reach as much as 30 feet in diameter, but from 10 to 20 feet is the most common size. The smaller sizes are heated by means of a steam jacket and the larger sizes by means of internal steam coils or pipes. At the top of the pan is a dome or large pipe connected with a " catch all," which serves to trap all liquid carried along mechanically with the steam and return it to the pan, while the steam passes to the condenser. The condenser may be a coil of piping surrounded with cold water or some special form of surface or jet condenser.

Fig. 92 shows a copper vacuum pan made by Blair, Campbell and McLean, Ltd., Glasgow, for concentrating sugar solutions, a branch of industry in which vacuum pans were used as far back as 1813.

This pan is 10 feet in diameter, and has a heating surface arranged in seamless copper coils, each coil having a steam inlet valve and pipe, pressure gauge, drain pipe, and steam trap.

FIG. 92.—COPPER VACUUM PAN.

FIG. 93.—CAST IRON CALANDRIA VACUUM PAN.

Fig. 93 shows a 12-foot cast-iron calandria vacuum pan
made by the same firm, in which the heating surface

9

consists of straight tubes of copper or brass expanded
into gun-metal or steel tube plates having a large circulating tube in the centre to assist the circulation.
The steam belt is placed in the cylindrical part of the
pan or effect, as it is called, and contains several hundred
vertical copper tubes about 5 feet long and 2 inches in
diameter.

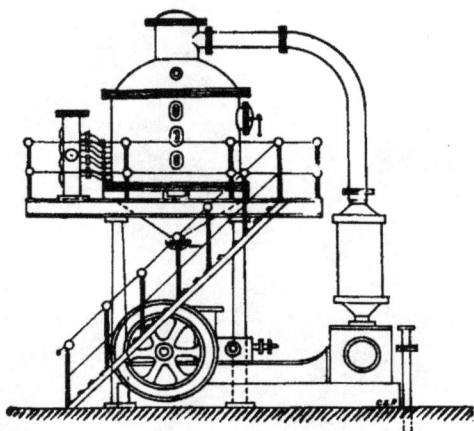

FIG. 94.—G.A. VACUUM PAN: JET CONDENSER.

Fig. 94 is a diagram showing the arrangement of a
vacuum pan having a jet condenser and wet vacuum
pump, and Fig. 95 is a similar diagram showing a vacuum
pan having a condenser with a barometric leg and
receiver, together with a dry vacuum pump.

There are many makes of vacuum pans; but in general
the method of operation is for either live or exhaust
steam to enter the coils or steam belt, usually, but not
always, at low pressure. Any water which condenses
here is drained away to the boilers or used for boiling
out or thrown away.

As the liquor in the pan boils, the vapour passes to a

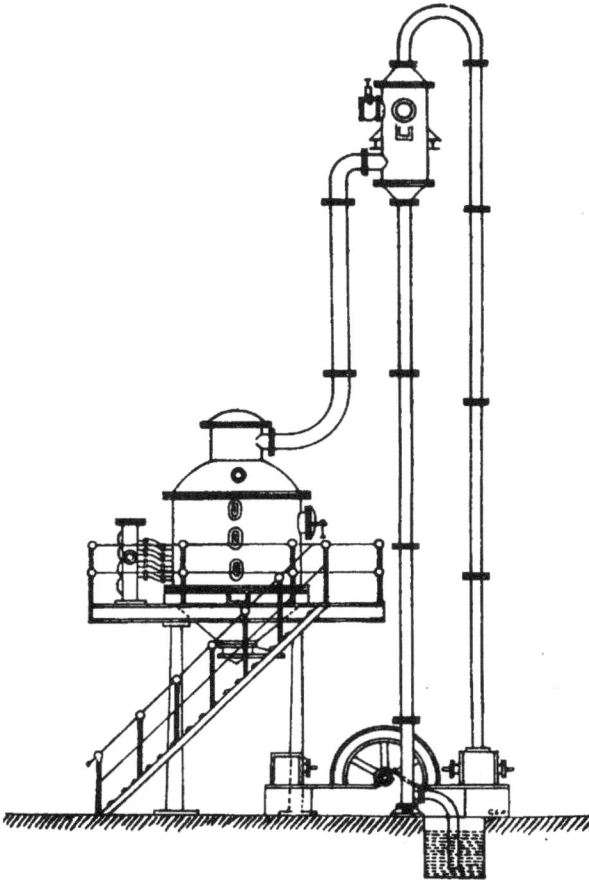

FIG. 95.—G.A. VACUUM PAN: TORRICELLIAN CONDENSER.

condenser, which by condensing the vapour helps to maintain the vacuum. Theoretically, the vacuum pro-

duced by the pump at the commencement of operations
should be maintained by the action of the condenser,
but in practice it is found necessary to work the pump

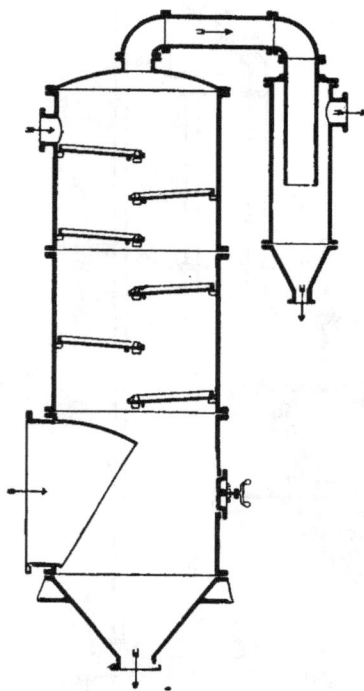

FIG. 96.—INJECTION CONDENSER:
SECTION.

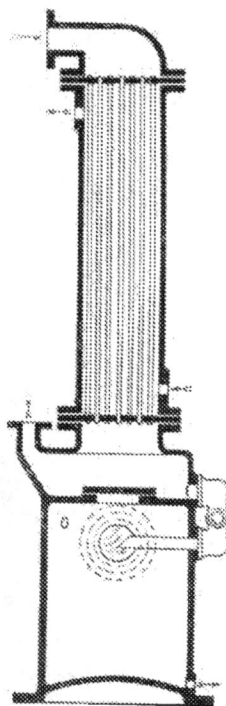

FIG. 97.—SURFACE CON-
DENSER: SECTION.

throughout the process to an extent depending upon the
efficiency of the particular condenser used. As a general
rule the water from the condenser is allowed to go to
waste, and as its temperature is low—about 80° to 90° F.

—not much heat is wasted, but in some cases this water may also be utilized.

In order to secure the highest efficiency, the vacuum must be as high as possible, and for this reason both the pump and the condenser must be as efficient as possible. It is essential that the condenser should have a sufficient cooling capacity to deal with the volume of vapour to be condensed. There are two types of condensers in common use—the jet type and the surface type, as shown diagrammatically in Figs. 96 and 97 respectively.

With the surface condenser the evaporated liquid is completely recovered and a pump of the dry type can be used, but with the jet condenser the vapour is mixed with the condensing water (equal to about forty times the weight of vapour condensed) and a larger pump of the wet type is required, but the operation is not so costly.

In the case of the surface condenser the receiver is usually directly connected, and arrangements are provided for discharge from time to time without interrupting the main process. Jet condensers are often connected with a barometric leg 35 feet high in which the condensing water stands and overflows into a well without breaking the vacuum.

Multiple-Effect Vacuum Pans.—As far back as 1830 the suggestion was made of connecting vacuum pans in series and regulating the pressure in each pan so that the work done by the steam could be greatly increased, but it took many years before the method was in general use.

In the multiple effect system each vacuum pan acts not only as an evaporator, but also as a boiler, producing heated vapour for boiling in the next pan and acting as a condenser for the preceding pan. Suppose, for example, steam at about 212° F. is admitted to the first pan of a series, and the vacuum maintained in that pan is 15 inches. The vapour from the first pan will have a temperature of

175° F., and passing into the second pan, which has a vacuum of about 24 inches, produces more vapour at 140° F., which in turn passes to the third pan, having a vacuum of 27 inches, and so on until the highest economical vacuum attainable limits the process.

The varied applications of the vacuum-pan process have produced a corresponding variety of makes, which differ considerably in detail and operation, but which would require more space than is available in this book for even the briefest examination.

Nearly a hundred years ago an eminent authority drew attention to the following points as constituting the essentials of a good evaporator: (1) The liquor to be evaporated as quickly as possible, so as to avoid any alteration; (2) the movement of the liquor to be accelerated by giving it great speed and spreading it out in the form of a thin film.

These are the principles involved in the design of the Kestner patent film evaporator, and hence show a great departure from those embodied in the older types of vacuum pan.

It is a truly continuous apparatus, the weak liquor being fed by gravity or by means of a pump at a constant rate into the bottom box of the evaporator, and the concentrated liquor at the predetermined density issuing continuously from the separator. The heating surface and the passage of the liquor through the tubes are so disposed that the liquor to be concentrated passes over the heating surface at high speed in the form of a thin film, whilst the steam space surrounding the tubes is so arranged that the outer surface of the tubes is swept at high speed by the hot steam, so that not only is the heat transfer improved, but the condensed steam is rapidly brushed off the heating surface, thus preventing any loss of efficiency from the tubes becoming water-logged. In addition, the arrangement is such that accumulation of air and non-condensable gases in the steam space is entirely prevented, and so the whole of the heating surface is able to exert the maximum efficiency. The advantage of the high velocity of the liquor and its short-time contact with the heating surface is threefold: The physical properties of delicate liquors remain absolutely undamaged, owing to the fact that they are only in momentary contact with the heating surface. Secondly, the rate of heat transfer is enormously increased, owing to the fact that the presence of large masses of water in feeble circulation is entirely eliminated. Thirdly, the formation of scale on the internal surfaces of the tubes is greatly reduced and in some cases prevented, giving a great range of application to the apparatus.

The design of the evaporator has a further advantage in that the ground space occupied is exceedingly small compared with that of many other pans, and that it lends itself to the subsequent addition of effects, so as to convert a single effect into a multiple, with the corresponding steam economy.

Fig. 98 shows a sectional view of a Kestner single-effect climbing film evaporator. It consists of two parts—namely, the calandria and the separator, the former being composed of a shell or casing containing the evaporating tubes, which are about 23 feet long and fixed into the upper and lower tube plates.

The liquor is fed into the apparatus at the lower inlet T, and passes from the feed box into the tubes. The steam or exhaust vapour, whichever medium may be employed to heat the liquor, enters the calandria of the evaporator at A. The liquor begins to boil in the tubes because they are surrounded by steam, and as ebullition takes place a column of vapour rises up the centre of the tube. This vapour travels at a high velocity, and at the same time draws up a film of liquor, which forms on the inner surface of the tube continuously without dry patches, so preventing any danger of burning any substances sensitive

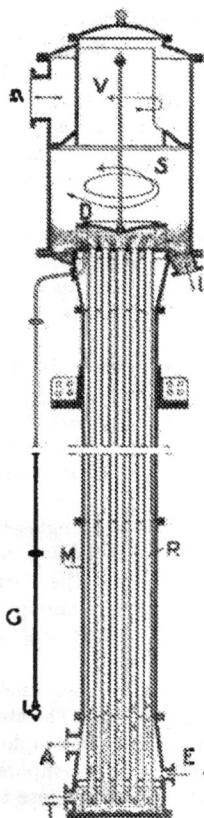

Fig. 98.—Kestner Climbing
Film Single-Effect Evap-
orator.

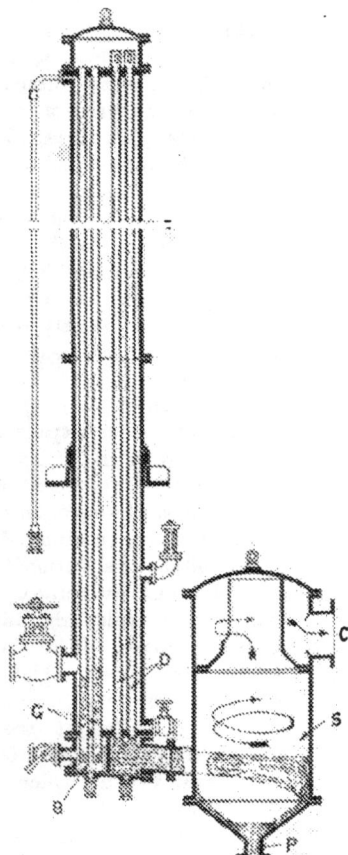

Fig. 99. — Kestner Falling
Film Single-Effect Evaporator.

to heat. Above the calandria is the separator S, which
consists of a cylindrical vessel containing a centrifugal

baffle placed immediately above the tubes, and so constructed that the liquor and vapour rising up the tubes in the calandria strike against the curved vanes of the baffle with such velocity that, due to centrifugal motion, there is complete separation of liquor and vapour. The concentrated liquor passes down the outlet L, and the vapour, passing after through the save-all, leaves the separator at B. The long shell of the Kestner gives two important advantages over the calandria of the ordinary type of vacuum pan: first, the distilled water can be removed easily by running it off at the opening E ; and secondly, all air and non-condensable vapours in the heating system are removed at G ; thus all the tube surface remains operative.

For such liquors as glue, gelatine, and the like, the final concentration is made in a falling film evaporator, of which a sectional view is shown in Fig. 99. The apparatus consists of—(1) Tubes 18 to 23 feet long secured in the upper and lower tube plates, divided into two groups G and D, forming the climbing film and falling film tubes respectively. (2) The separator, placed below the lower tube plate, contains the feed box, and is fitted with a centrifugal baffle and the necessary openings for the concentrated liquor outlet and for the discharge of the vapour. The liquor to be concentrated is delivered into the feed box B, from which it passes into the tubes G, where the climbing film action takes place. The liquor and the vapour arrive in the upper box above the tube plate, where they are distributed to the tubes D, the liquor running down as a thin film, and the high-speed vapour forming a core in the centre of the tube. Both liquor and vapour pass into the separator S, where by means of centrifugal action complete separation takes place, so that the vapour passes at C and the liquor at P.

By means of this apparatus substances such as liquorice, gelatines, glue, dyestuffs, milk, fruit juice, and sugar can be delivered in such a high state of concentration that the extract solidifying on cooling can be run into drums or moulds ready for transit.   Many substances can be concentrated without a vacuum in this

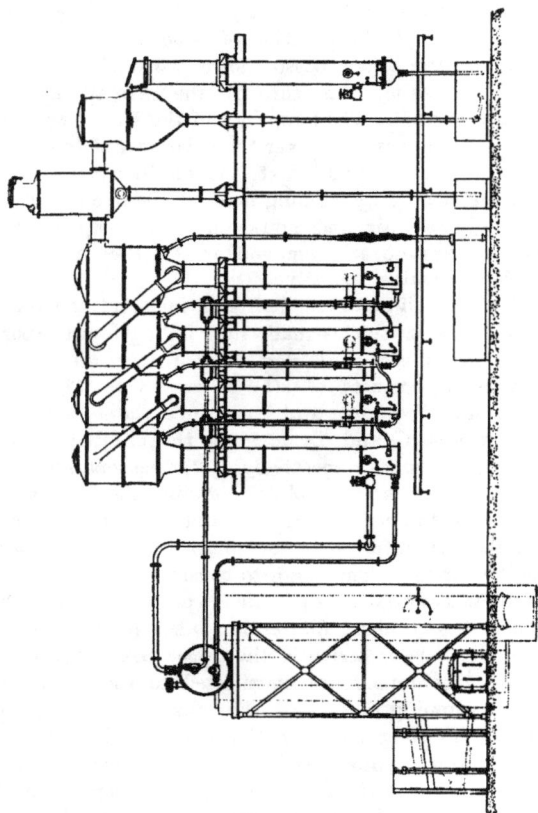

FIG. 100.—KESTNER QUADRUPLE-EFFECT, ETC.: DIAGRAM.

apparatus‚which require a vacuum in the older types of apparatus.   By this method all the steam required for the auxiliaries is saved, and when multiple effects are

used great economy is obtained, as only a single feed
pump is required; moreover, where cooling water is
scarce, the possibility of avoiding the use of a condenser
is a great advantage.

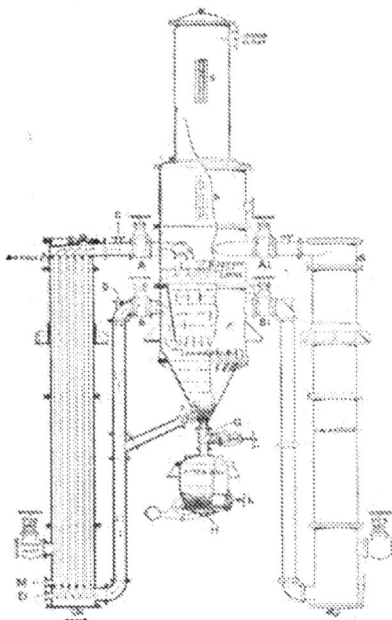

FIG. 101.—KESTNER "SALTING" TYPE EVAPORATOR: SECTION.

Fig. 100 shows a direct-fired evaporator followed by a
quadruple effect, and a single-effect finisher which is
stated to produce caustic liquor of 60 per cent. Na_2O.

Fig. 101 shows a section of a " salting " type evaporator
which is used for the concentration of dual solutions
and for the production of crystals direct in the separator
instead of in the usual trays. Separation of salts by

crystallization, owing to difference of solubility, can be effected in this apparatus—e.g., mixtures of NaCl and NaNO₃, NaOH and NaCl, NH₄NO₃ and Na₂SO₄.

The separator is a cylindrical vessel round which several calandrias can be grouped. Liquor is admitted to the separator to above the opening B, at which point the calandria can operate, the height being checked through the sight glasses S. Steam is turned on in the calandria, the liquor passing through B down into the bottom box of the calandria through the tubes, and back into the separator at A. The circulation is continuous, and the crystals formed are by means of the circular baffle deposited on the bottom of the cone, whilst the liquor passes through B and the vapour formed passes into the save-all, for further treatment as desired. At the bottom of the separator is a valve capable of dealing with liquors heavily charged with salt crystals, and discharging into a salt box or filter box, which latter has a bottom cover arranged with a balanced hinge so that the filtering medium can be easily inspected.

By means of the isolating valves A and B, any one of the heating units can be shut off from the system for cleaning or repairs without interfering with the running of the rest of the apparatus.

With multiple-effect apparatus there is an economy of steam and condensing water amounting to one-half in a double effect and two-thirds in a triple effect of that consumed in a single-effect apparatus. Multiple-effect evaporators have not been universally employed because materials which are sensitive to high temperatures could not be evaporated in such an apparatus without suffering injury, as the operation of a multiple effect requires evaporation temperatures up to about 176° F. in a triple effect, and still higher for a larger number of effects. For small and medium quantities of liquid, and where work is done for an hour or two at a time, the use of a multiple effect was precluded because of the difficulty of starting and stopping it, the large space required, and the very considerable first cost.

The " Multiplex " evaporator made by Blair, Campbell and McLean, Ltd., Glasgow, overcomes some of these objections, although high temperatures in the first effect cannot be avoided. By special construction the quantity of liquid contained in each effect is so small that the liquid remains in it only one or two minutes at the utmost, and is then drawn off into the next compartment, which is at a lower temperature, and finally, after a short time, is completely concentrated and discharged from the apparatus.

Fig. 102 shows a view, and Fig. 103 shows a section of a " Multiplex " triple-effect apparatus. When the apparatus has been exhausted of air the liquid to be concentrated enters at 1, into the double bottom 2, and rises to a uniform height in the tubes 3, which are surrounded with steam which enters at 4. The liquid very soon boils, bubbles of steam first forming at the underpart of the heating tubes and increasing to a stream of high velocity. The liquid attaches itself to the tube walls, and the steam drives it high up these, so that the middle and upper parts of the tubes are no longer filled with liquid, only the tube walls are wet with it. The bubbles of froth formed in the lower part of the tubes are thus broken up, so that the liquid leaves the tubes in the form of drops which spring off the top edges of the tubes. Close above the upper tube plate another plate is fixed, between which the velocity of the steam is so great that the drops have no time to sink down, but are blown towards the tubular connection 5, and through it into the separator 6. There the liquid sinks to the floor, and thereafter it rises through the tube 7 into the double bottom 2, and the heating tubes of the second effect, whilst the vapour rises into the heating compartment 8, and there serves as heating steam. The process of evaporation is repeated there in the same way

FIG. 102.—" MULTIPLEX " FILM TRIPLE-EFFECT EVAPORATOR

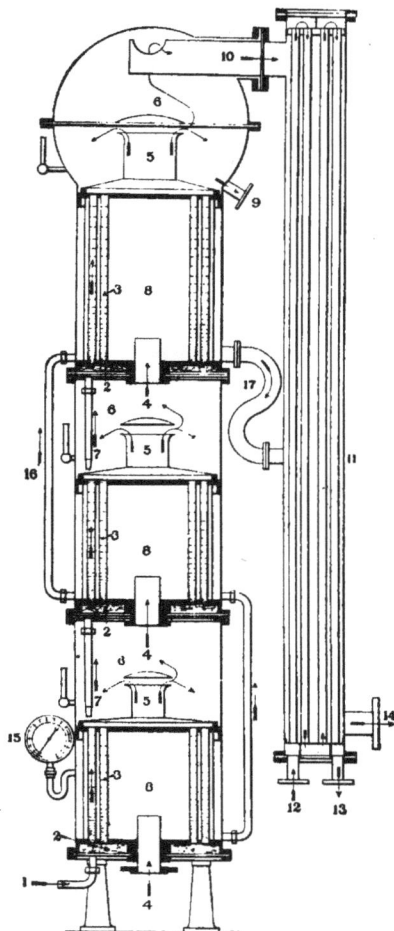

FIG 108.—"MULTIPLEX" TRIPLE-EFFECT EVAPORATOR: SECTION.

as in the first effect, and again in the third effect, from which the liquid is drawn off at 9 after it has been concentrated to the desired degree. The vapour from the last effect is then condensed in the condenser through pipe 10.

Frothing interferes very much with the evaporation in a vacuum apparatus, but in this case the liquid cannot rush through the tubes in the form of froth, but adheres to the walls and evaporates. The central entrance of the steam into the heating space causes an energetic circulation, so that all the heating tubes are heated with the utmost possible uniformity, and no dead corners are formed in which material that would generate deleterious gases can collect. The apparatus is constructed in such a way that the whole amount of liquid contained per square foot of heating surface is only a few gallons. Every particle of the liquid, therefore, only remains a few minutes in the apparatus, and is then forced out by the liquid following upon it, so that the material undergoing concentration is very soon withdrawn from any deleterious action of high temperatures.

The mode of operation is very simple, and consists of putting the vacuum pump in action and setting the valves. It sucks in the liquid which is to be concentrated through a pipe, and ejects it in the concentrated state through another pipe, the degree of concentration being regulated by the amount of the feed.

The sugar and the salt industries have had the longest connection with the vacuum-pan process, and as both these industries are widely represented on the American continent, it follows that there are many types of pans made by American firms. Among the most important vacuum pans used in Canada and America are the Manistee, Lillie, Brecht, Craney, Oscar Krenx, Swenson, Sanborn, Wheeler, and Zaremba vacuum pans, each of which is worth consideration by the chemical engineer.

CHAPTER V

DISTILLING APPARATUS

DISTILLATION is the term usually applied to the application of the process of evaporation to the separation of a solution into its components. There are three principal parts which all types of distilling apparatus have in common—viz., (1) a vessel or still in which the material is heated; (2) a cooling apparatus for condensing the evaporated material; and (3) a receiver for collecting the condensed material or distillate.

The process of heating materials and collecting the bodies formed by the action of heat is termed destructive distillation, and, when the main object of the process is the residue in the still, so that this part of the apparatus undergoes great modification, the terms roasting, burning, glowing, and firing are used, and the apparatus termed a muffle, furnace, kiln, etc., as the case may be.

It will be obvious from what has already been said about vacuum pans that they are a form of still, and, indeed, they are often used for this purpose in suitable cases, but the separation of liquids of close boiling point demands the type of apparatus usually termed a still.

The Column Still.—This commonly used apparatus, of which a diagrammatic view is shown in Fig. 104, derives its name from the column or dephlegmator A, through which the vapours from the boiler are made to pass before being condensed and collected in the receiver. This column, which is fixed over the still or boiler, contains a number of shallow cups or plates placed at intervals, thus dividing the column into a series of chambers between which communication is maintained by means of perforations in the plates or by small tubes so arranged that a small quantity of liquid can be retained in each cup.

When the process has been continued for a sufficient time for the column to take up a steady state there is a definite drop in temperature from the top of the column

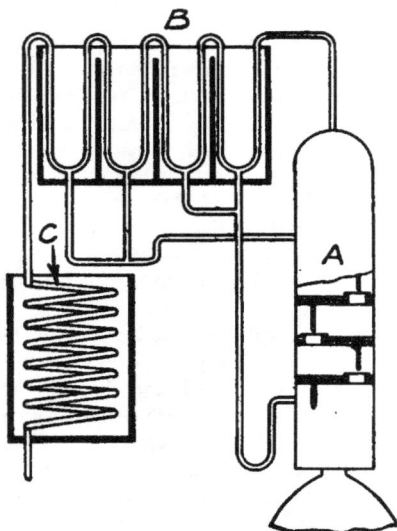

FIG. 104.—DIAGRAM OF STILL COLUMN.

downwards, and each cup contains a small quantity of liquid with a correspondingly higher boiling point through which the oncoming vapour from the boiler must bubble. In each, the partial pressure of the constituents depends upon the temperature and the composition of the liquid therein, so that the higher boiling constituents, as they fall back down the column, enrich themselves at the expense of the ascending vapours, until only the lower boiling or more volatile bodies issue from the top of the column. From the top of the column the vapours pass through a series of U-tubes B, which are surrounded by a bath kept at a definite temperature. From the bottom of these U-tubes draining pipes lead back to the column, and are so arranged that they discharge into it at various points, depending upon the boiling point of the condensate, progressively from the bottom of the column with the highest boiling point to the top with the lowest.

By this means the vapour which issues from the U-tubes and is condensed in the coils C is comparatively free from foreign substances, and has a constant boiling point. By continuing the process this constituent may be generally entirely recovered, and the apparatus taking up a fresh steady state, the next higher boiling constituent may be recovered, and so on.

This is the method known as fractional distillation, · and it should be noted that it is not applicable to all types of liquid mixtures, of which a short account is given in the Appendix.

Fig. 105 shows a patent rectifying still made by John Dore and Co., London. It is designed for the strengthening of weak alcoholic liquors and the recovery and purification of solvents, such as ether, acetone, and the like, and it is claimed that in one operation it will produce alcohol of 0·815 specific gravity from weak liquors of about 0·967 specific gravity.

Fig. 106 is a diagram illustrating a continuous still made by George Adlam and Son, Ltd., Bristol. This still consists of—

1. A boiling or analyzing column made in sections of cast iron and fitted with copper plates between each joint, on which are fitted copper bell plates and dip pipes. In the bottom chamber is placed a steam heating coil for heating the liquor.

2. A rectifying column constructed of strong copper and made in flanged sections. This is also fitted with copper bell plates, bells and dip pipes similar to those

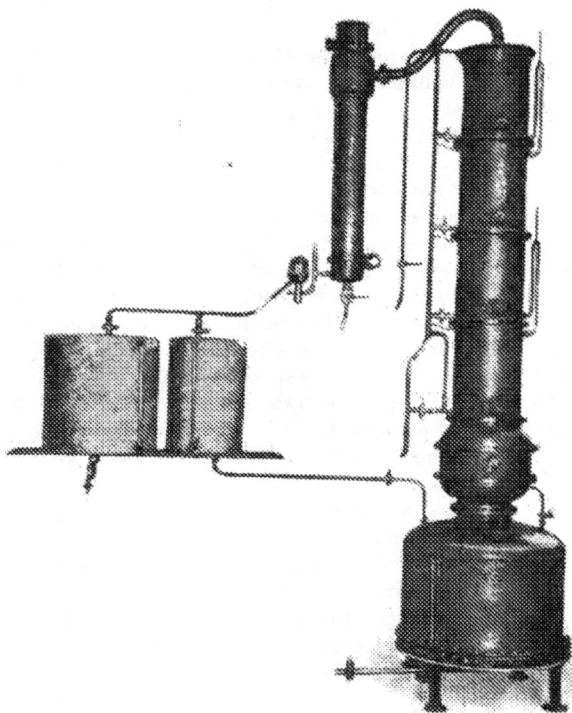

FIG. 105.—RECTIFYING STILL.

in the analyzing column, but with four plates to each section.

3. A rectifier or reflux condenser.
4. A condenser with tube plates and tubes.
5. Sight glasses or still watchers.

The liquor is fed into the still at the seventh section of the analyzing column, and exhausted, waste, or spent liquor is run off from the bottom chamber.

The vapours pass from the analyzing column to the rectifying column through the connecting pipe, and

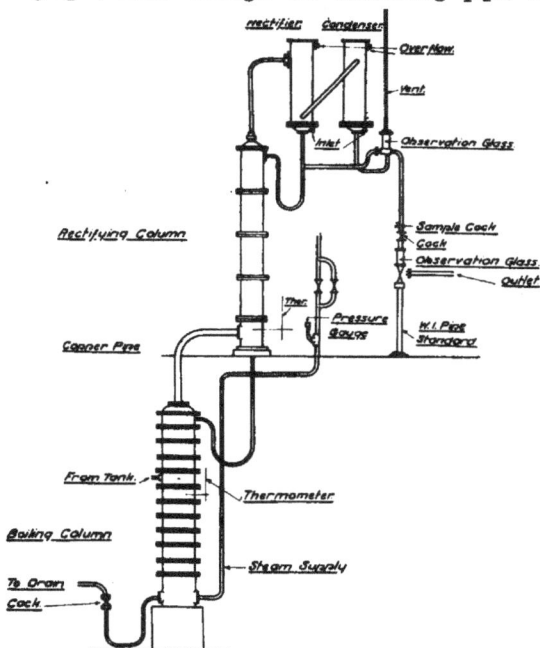

FIG. 106.—CONTINUOUS DISTILLATION APPARATUS: DIAGRAM.

from the rectifying column to the rectifier through the bent pipe shown. Portions condensed here are returned to the rectifying column through a draining pipe at the bottom, and the rest of the vapour passes on for condensation in the spirit condenser, where it can be observed and discharged as required.

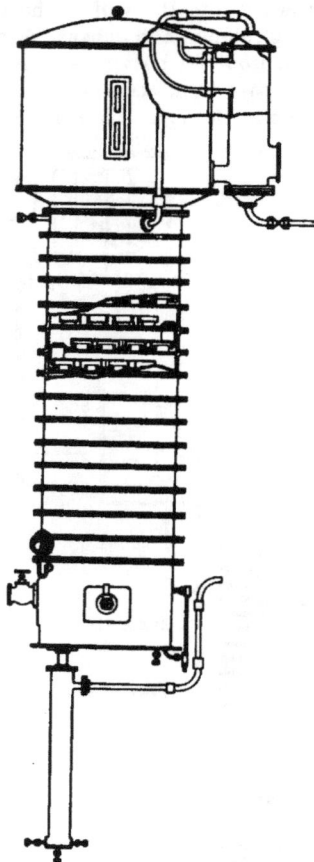

FIG. 107.—CONTINUOUS STILL: DIAGRAM.

Fig. 107 shows a continuous still made by George Scott and Son, Ltd., London.

The Coffey Still.—This particular type of still, named after the original maker, Æneas Coffey, whose successors are John Dore and Co., London, has found a most varied application in many of the branches of chemical industry. To trace out the application of the principles of the Coffey still to the various branches of industry is impossible for the purposes of this book, but the student should make himself acquainted with the main outlines of the process, and always be on the look-out for its practical application in the industrial world.

Fig. 108 is a diagram which gives a rough idea of the principal parts of a Coffey still. It consists of two columns

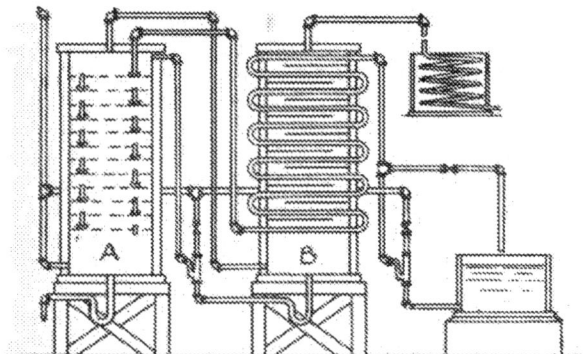

FIG. 108.—DIAGRAM OF COFFEY STILL.

or towers *A* and *B*, known as the analyzer and rectifier respectively. The internal arrangement of the analyzer is similar to that of a column or dephlegmator on a large scale. The rectifier consists of a column which contains a tubular coil, through which the liquor is pumped and discharged at the top of the analyzer over the perforated plates or trays. Steam or some other suitable vapour is admitted at the bottom of the analyzer, and, rising through the descending stream of liquor, is partially condensed whilst vaporizing the volatile constituents of the liquor. The action of this tower is exactly similar to the action of the column of an ordinary still, so that from the top of the analyzer a pipe leads away a heated mixture of steam and vapour of the volatile portions of the liquor. This hot vapour is in turn delivered to the bottom of the rectifier, where it gradually rises through the coils of the liquor tube, heating the contents and at the same time condensing out the higher boiling portions, including the steam, which run down to the bottom. The most volatile portions pass out of the rectifier at the top, and are led to a condensing apparatus and receiver. The liquid which collects at the bottom of the rectifier is pumped up to the top of the analyzer, and there discharged over the plates, together with the liquor from the tank. With proper working, by the time the liquor has reached the bottom of the analyzer all the volatile portion has been extracted, and it may be drawn off and discharged as spent liquor. It will be noticed that the apparatus practically consists of two columns arranged on the counterflow system.

It need hardly be mentioned that the success of the process depends both upon the design of the parts and also upon the experience and skill of the operator.

Among the many applications of this apparatus a very interesting one is its use in the liquefied gases industry for the separation of the constituents of mixtures such as liquid air, liquid natural gas, etc., whereby formerly rare gases, such as argon, neon, and helium, are obtained in commercial quantities.

Extraction Plant.—One of the most interesting modifications of distilling apparatus is found in its application to the extraction of oils and drugs.

Fig. 109 shows a continuous extraction apparatus made by John Dore and Co., London, which is designed for the extraction of drugs by means of alcohol, ether, acetone, petrol, benzene, etc., and is so arranged that there is little or no loss of the solvent used. It consists of three vessels—viz., an evaporator, extractor, and condenser,

FIG. 109.—EXTRACTION APPARATUS.

all mounted on a self-contained stand with the necessary cocks and connections. The method of operation is for the extractor to be filled with the plants or roots to be treated, while the evaporator is filled with the solvent. This latter vessel, heated by a steam jacket, vaporizes the solvent, which is conveyed to the overhead condenser, where it is condensed and allowed to flow or percolate through the material in the extractor and back into the evaporator again, together with the dissolved substances. Here the solvent is again vaporized and returned for extraction, and so the process is carried on until the material in the extractor is completely extracted. At this stage the solvent is again vaporized and drawn off from the condenser, while the product is collected from the lower vessel as required. The centre extraction vessel is also fitted with a steam jacket which can be used for driving off any solvent remaining in the exhausted mass, or for heating the drug during the process of extraction. Both vessels are fitted with removable covers for cleaning purposes and for charging.

Fig. 110 is a diagram showing the arrangement of an oil extraction plant made by George Scott and Son, Ltd., London.

The distillation of such substances as crude petroleum and coal tar involves both distillation proper and destructive distillation. The plant used is comparatively simple in nature, although in most cases of huge size. For the fractional distillation of crude petroleum, cylindrical steel shells up to a size of 15 feet in diameter and 42 feet in length are set horizontally in brickwork, leaving the upper half exposed except for an iron cover. Of the two methods of firing—end firing and side firing—the latter is preferred on account of the greater control of the still which ensues. The stills are fitted with the usual dome, from which the vapour main of 12 inches to 18 inches in diameter leads to the condensers, which are very often simple coils immersed in a water bath.

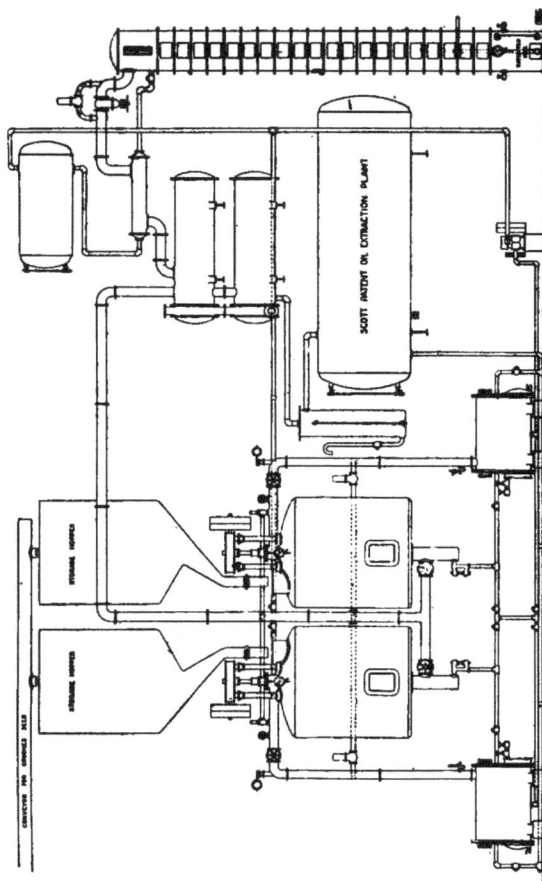

Fig. 110.—Scott Oil Extraction Apparatus: Diagram.

In this type of still the various fractions are collected until about 10 per cent. of the original oil is left as a tarry residue, which is removed and distilled in tar stills.

Sometimes the distillation of the oil is carried on till the residue is destructively distilled to coke. In this

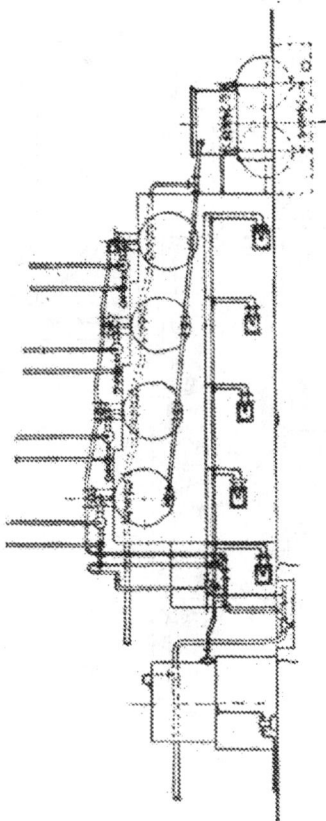

FIG. 111.—LUBRICATING OIL DISTILLING PLANT: ELEVATION

case the vapours are led through a kind of column, which takes the form of a number of towers each of which corresponds to a chamber of the smaller column. These

towers consist of two chambers connected by tubes,
around which the air circulates and in which the vapours
are condensed and run down into the bottom chamber,
whence they are drawn off as a separate fraction after
passing through water coolers.   The towers are connected
to the still in series, the vapour entering at the bottom
of the first and passing out at the top to the bottom

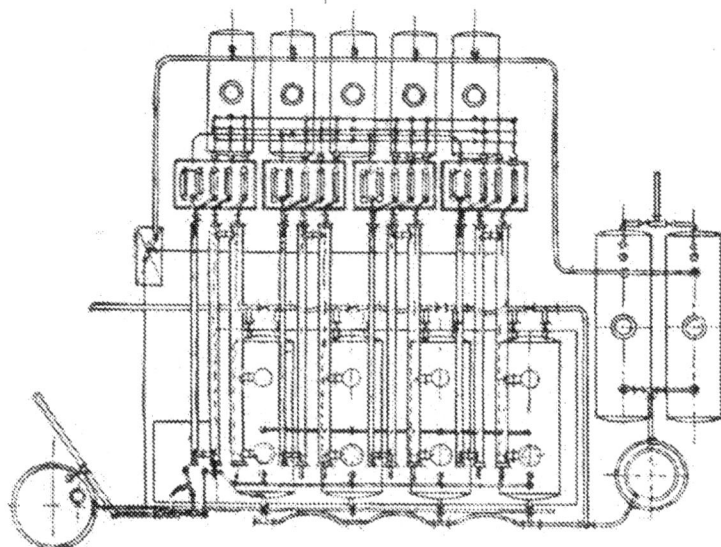

FIG. 112.—LUBRICATING OIL DISTILLING PLANT: PLAN.

of the second, and so on.   In certain cases steam is
blown into the still, which has the effect not only of keeping
the mass agitated and preventing overheating of the
bottom portion, but also, by its additional pressure on
the surface, lowers the partial pressure necessary for any
constituent to boil, and so causes that substance to boil
off at a lower temperature.   It need hardly be mentioned

that some manufacturers obtain the same effect by working the process under a vacuum.

On account of the fact that a high distilling temperature is injurious to the product, in the case of lubricating oils, the process of vacuum distillation is in common use.

Figs. 111 and 112 give a view in elevation and plan of a continuous vacuum oil distilling plant for lubricating and paraffin oils made by W. J. Fraser and Co., Ltd., Dagenham, Essex.

The plant is so designed that a high vacuum is maintained in the entire system by means of a vacuum pump under a continuous or periodical distillation, the distillates being collected in their respective receivers. This plant may be advantageously connected direct with the crude oil distillation plant, effecting thereby a further saving in fuel and labour. Among the advantages of this type of plant are (1) a high quality product due to low temperature and high vacuum; (2) no cracking, as the vapours do not have contact with highly heated plates in the still.

For the distillation of tar, similar stills having a capacity of about 5,000 gallons are in common use, but very often the still is of the vertical type, having a convex top and concave bottom. Constructed of $\frac{1}{2}$-inch boiler plate with a bottom of $1\frac{1}{2}$ inches set in a brick arch over the fire, the lower half is heated by the hot gases from the fire being made to circulate round it by means of flues. The vapours are led away to the usual type of condensing coil, and provision is made for running off the pitch from the still into a vessel for cooling.

Retorts.—The greater part of the labours of the early chemists was devoted to the heating of all manner of materials in an alembic or retort and investigating the nature of the resulting products. This form of distilling apparatus in some cases yielded important results of commercial value, so that industries were started and the retort was developed in accordance with

the particular needs.   The term "retort" is now generally
used to indicate that part of the apparatus in which
the heating of the material is carried on.   The construction of a retort depends entirely upon the particular
industry in which it is used and the requirements of the
individual manufacturer.

In the nitric acid industry a retort is required in which
sulphuric acid and nitrate of soda can be mixed and
heated, resulting gases collected, and provision made
for removing the nitre cake.

These retorts are usually cast-iron cylinders about 5 feet
in diameter and 10 feet in length, closed at either end by
stone or cast-iron plates, one of which is pierced for the
separate feeding of the acid and the soda, and the other
for the exit of the gases and the discharge of the nitre
cake.   These retorts are fixed in a brickwork setting and
only require a comparatively small fire area.

Another common form is known as the pot still, which
consists of a pot made up of three sections luted together
with an acid-resisting cement.   The upper sections are
lined with bricks, but the bottom section is unlined, as
it is not so liable to corrosion, and to allow of easy heat
transfer.   The bottom section is provided with an outlet
for the discharge of the nitre cake, and the top section is
provided with a charging door and gas exit tube.   As a
rule the pots receive a charge of about 1 ton of material
a day, which is gradually distilled.

The retorts used in a by-product coke oven are long
narrow structures of firebrick about 30 feet long, 6 feet
high, and 1½ feet wide, arranged side by side, separated
by flues.   The ends are closed by sliding iron doors,
which are luted during operations, and which can be
raised at the end of a run and the whole of the contents
pushed out by mechanical means.

The retorts used in the coal-gas industry are of three
kinds—viz., (1) horizontal, (2) inclined, and (3) vertical.
There is considerable variation in the length and

cross-section of these retorts and in the method of heating, although the use of producer gas is growing in favour. Horizontal retorts are usually provided with mechanical stokers, and are charged to about two-thirds of their capacity. Inclined retorts are charged by feeding in at the top, and discharged from a door at the bottom, thus saving a certain amount of labour. The gas is drawn off at the bottom of the retort, but it is said that the yield is smaller than in the case of other types of retorts. Vertical retorts are arranged in groups for filling at the same time, and the gas is drawn off at the top, while the coke is removed from the bottom and used directly for making producer gas for heating the retorts.

An exceedingly important modification of the gas retort is that designed by Mr. Dowson for the production of a cheap gas fuel for driving gas engines and for heating work of all kinds where cocks and burners are used.

Briefly, the gas is made by passing superheated steam, mixed with air, through red-hot fuel in a vertical gas producer. The steam is decomposed, the oxygen combining readily with the carbon of the fuel, and the combustible constituents of the gas consist of hydrogen, carbon monoxide, and a small percentage of marsh gas. The process is continuous and automatic, and there is no outside fire, as there is with an ordinary retort; the cost of repairs is low, and the apparatus is simple and easy to work. The gas is made as quickly as it can be consumed, and its production being governed automatically to suit a varying rate of consumption, it can be stopped completely for meal-times or when laying off. The gas is cooled, washed and scrubbed, and passed into a gasholder when required, although in many cases the latter operation is found not to be necessary.

The original Dowson plant is worked with a jet of steam at pressure, acting as an air injector, and is known as the pressure plant, but in the more recent plant the suction plant, air and steam are drawn in by means of a fan or gas engine, the only difference in the result being that pressure gas has a little higher calorific power and is more useful for heating work than the suction gas.

Both types are worked with anthracite (peas, beans, or nuts), charcoal, or gas coke, which latter should be in pieces of ½ to ¾ inch cube, and should not contain more than 10 to 12 per cent. of ash. Owing to the formation of tar, special plants are needed for using bituminous coal and for the utilization of wood refuse, shavings, sawdust, etc.

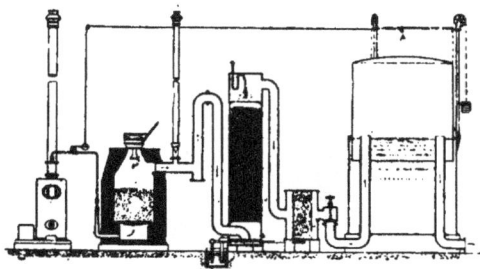

FIG. 113.—" DOWSON " STEAM JET PRESSURE GAS PLANT.

Fig. 113 shows a diagrammatic sectional view of a Dowson pressure plant. A jet of steam at pressure from a small independent boiler, or from a factory or other boiler near the gas plant, plays in an open air pipe, and the mixture of steam and air is forced into the fire in the producer, the fuel being put in through the hopper on the top. The gas is made continuously so long as the jet of steam is working, and if shut off the production of gas ceases at once. On the steam pipe there is a governing valve and lever actuated by the rise and fall of the gasholder, so that the rate of production is governed automatically to suit a varying rate of consumption.

11

After the gas leaves the producer it passes through a water seal, and then through coke and sawdust scrubbers.

The consumption of anthracite or charcoal is about 13 pounds, or of coke about 14 pounds per 1,000 cubic feet of gas, and it may be taken for the purposes of costs comparison that 4,000 cubic feet of this gas are equivalent to 1,000 cubic feet of town gas.

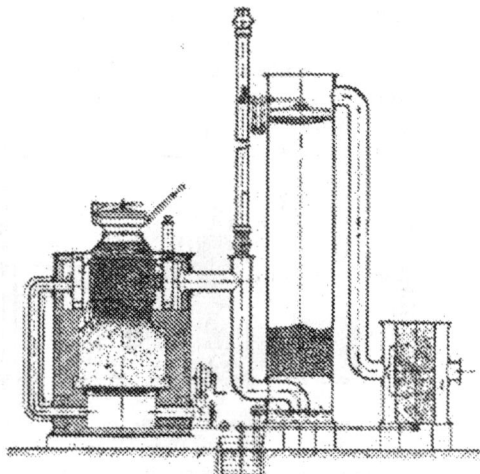

FIG. 114.—" DOWSON " SUCTION GAS PLANT.

This type of plant is suitable when there are two or more gas engines, when there are engines and heating work, or when there is heating work only. The gas mains are then simplified, and it is also more easy to start two or three engines from a pressure plant than from a suction plant.

When this type of plant is used for engine work the consumption of anthracite or charcoal of average quality is about 1 pound per b.h.p. hour, the actual consumption depending somewhat on the efficiency of the engine. With coke the consumption is a little higher.

Fig. 114 gives a sectional view of the Dowson suction plant. In this case the steam is formed in a vaporizer inside the producer, near the top, and the steam and air are drawn into the fire at the bottom by means of a fan or by the suction of the engine, which works in combination with the plant. Every plant has a small fan for blowing up the fire at the start, and when the engine is started this fan is stopped and the engine itself governs the rate of producing the gas to suit its own varying consumption. After the gas leaves the producer it passes through a water seal, and then through coke and saw-dust, as in the pressure plant.

Fig. 115 is taken from a photograph of a 30-h.p. plant. The chemical process of making the gas is the same as in the pressure plant, but as there is no inde-pendent boiler, no allowance need be made for raising the steam required, so that with a good engine the consumption of anthracite or average charcoal is about $\frac{3}{4}$ pound per b.h.p. hour. From tests which were made on a 40-h.p. plant the heat efficiency was found to be as high as 90 per cent.

For plants of about 200 h.p. and upwards it is found that bituminous coal is cheaper than anthracite, and so a special type of plant is used.

Fig. 116 gives a sectional view of a Dowson bituminous plant for making gas without tar. The special feature of the producer is that it is double acting—*i.e.*, air is drawn in through the top and through the bottom of the fuel column, as indicated by the arrows. The producer is open at the top, and coal is put in there, but there is no escape of smoke as air is drawn inwards by an exhaust fan. The upper part of the fire burns down-wards, the hydrocarbons are distilled off, and the coke which remains sinks downwards into the lower part of the producer, where it meets an upward current of steam and air and is converted into ordinary producer gas. The mixture of gases leaves the producer through

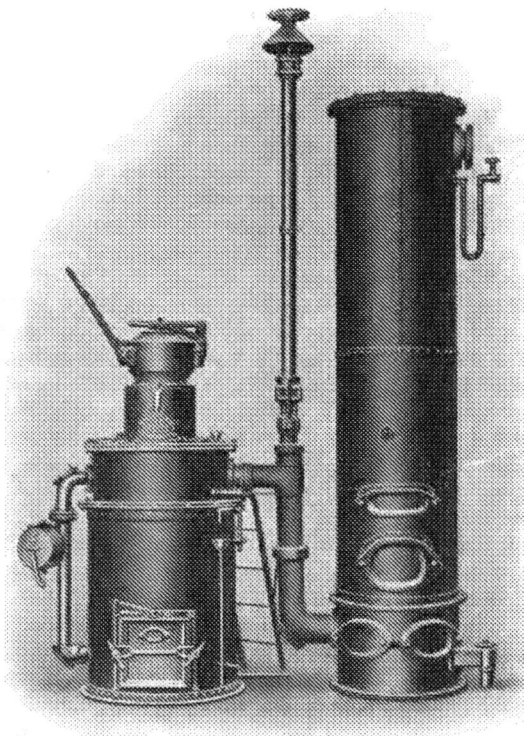

FIG. 115.—30 H.P. SUCTION GAS PLANT.

an outlet about halfway between the top and the bottom. The producer has a water bottom, so that clinker and ash can be drawn out while the plant is working, and almost any kind of coal can be used which does not contain more than 31 to 35 per cent. of volatile matter.   After leaving the producer the hot gas passes through a vaporizer to cool and also assist to raise the steam required.   It then passes through special scrubbers to remove dust, scot, etc.,

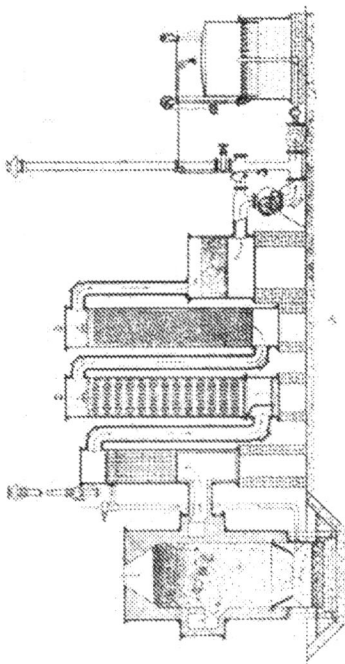

Fig. 116.—"Dowson" Bituminous Plant

but in this process there is no tar, as it is converted into gas in the producer, and no mechanical or other tar extractor is required.

The calorific value of this gas is nearly the same as that made from anthracite, and under good conditions the consumption of coal of fairly good quality in pieces of about ½ to 1 inch cube is a little over 1 pound per b.h.p. hour.

During the last few years the great development of the oil-hardening industry has created a demand for large quantities of pure hydrogen. In the Lane process this gas is produced by means of a special retort which is the result of the research work of Mr. Howard Lane during the past fourteen years. The retorts used are of the vertical type, and consist of cast-iron tubes 1¼ inches thick, 9 inches internal diameter, and 9 feet 9 inches long, having end covers for charging and discharging, and arranged in a brickwork casing. The basis of the process is the alternate oxidation and reduction of iron by steam and water gas respectively, and the purification of the hydrogen formed. The retorts are first charged with spathic iron ore, which on heating parts with its carbon dioxide and yields ferrous oxide. This oxide is then reduced by heating in a stream of purified town gas or water gas, and then subjected to the action of steam, whereby the iron is oxidized and hydrogen liberated. Although the process is chemically simple, the successful results obtained depend largely upon the inventor's mode of working. Since the production process takes twice as long as the oxidation process, Mr. Lane arranges three groups of retorts, so that two groups are reducing while one is oxidizing. In the experimental plant at Ashford the control valves are operated every 10 minutes, so that each retort produces hydrogen for 10 minutes every half-hour. From time to time it is found necessary to burn out the iron in a current of air, in order to restore its activity, the iron becoming poisoned by the accumulation of sulphur and other impurities which find their way past the scrubbers. Owing to the conditions of the reaction, an excess of water gas is needed to obtain complete reduction; hence a certain amount of this gas passes from the retorts unused, but at a later stage it is dried and used for firing the retorts. The purity of the hydrogen produced by this process is stated to be from 99 to 99½ per cent., and the cost, depending upon local conditions, is low enough for its production on a commercial scale.

Kilns.—This type of apparatus is used when it is necessary to subject material to the action of a high temperature in order to drive off moisture or some volatile constituent. They are mostly used in the cement and gypsum industries, and may be divided into the stationary and rotary types. Of the former type the primitive limekiln needs only a mention, but a modified form consists of a vertical steel cylinder lined with firebrick up to 10 feet in diameter and 50 feet in height. The fuel is kept apart from the limestone in two fireplaces built in the sides, and so arranged that the hot gases pass through the kiln and the ashes fall into a separate ashpit below.

In the chamber type of kiln a series of chambers are built round a central stack and connected to it by flues. The chambers, which are alternately charged with fuel and limestone, are so arranged that any one may be disconnected from the flue and separated from the other chambers by partitions as required. Thus the lime may be removed and the chamber recharged and set into operation with considerable saving of fuel. In the gypsum industry the kiln takes the form of a beehive with a flat floor resting on a cylindrical base in which are doors, each opening into a furnace. The kiln, which is built of brick, is about 16 feet high and 30 feet in diameter, and is arranged so that the hot gases are led through flues on the inner side of the kiln down through the material to an underground flue to the stack. As a rule a white heat is maintained for three days, when the lumps are removed and reduced to a fine powder for the purposes of cement.

The rotary calciner (Fig. 117) used in the gypsum

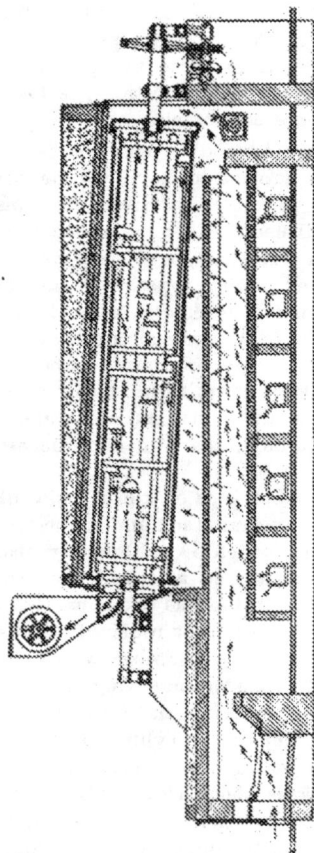

FIG. 117.—ROTARY CALCINER.

industry consists of an inclined cylinder 30 to 70 feet long and 5 feet or more in diameter, set at a small angle to the horizontal and caused to revolve slowly, having roller bearings and trunnions and a heavy geared driving wheel at one end. The cylinder is housed in a brick casing, in one part of which is situated the furnace, the bottom part consisting of chambers with perforated tops, about 2 feet above which is a perforated arch. Through these perforations cool air passes, and mixes with the hot gases, which are drawn by a fan connected to the top end of the cylinder through the bottom chamber and up the cylinder as the material passes down. A certain amount of the gases passes through the arch into the cylinder through ducts arranged in the length of the cylinder and protected on the inside to prevent any loss of material. Lifting blades running the entire length of the cylinder keep the material in motion during the ten minutes or so that it takes to travel the whole length.

By the use of a recording thermometer at the outlet and the operation of the cool air damper a steady temperature can be maintained throughout the operation.

In the Portland cement industry rotary kilns are used up to 150 feet in length, made of $\frac{1}{2}$-inch steel plates with single strap butt joints and lined with some refractory material. The cylinder, which has an inclination of about 1 in 15, is driven near its middle by a train of gears at a speed of from 25 to 55 revolutions an hour. The top of the kiln, where there is a water-cooled feeding device, projects into a flue connected with a firebrick-lined shaft provided with a door or damper. The lower end of the kiln has a removable firebrick cover having openings for the discharge of clinker and for the heating apparatus, which may consist of a jet of powdered coal, worked by a fan or compressor, which partly supplies the air necessary for combustion.

The Muffle Furnace.—When it is necessary to calcine material without having contact with the hot gases the muffle furnace is employed. The muffle itself is usually of firebrick, and the flues are arranged so that the hot gases first pass beneath the bottom of the muffle and then over the top back to a point near the grate, and thence to the chimney. In cases where any gas has to be discharged from the muffle a pipe is fixed to the top to allow of its ready escape.

The Reverberatory Furnace.—In this type of furnace, which has extensive application, the material treated is exposed to the direct action of the gases from the fire. It consists of an arched brick chamber lined with fire-brick, at one end of which is placed a grate for heating, and at the other end a chimney to carry off the waste gases. The material is placed on the floor of the arched

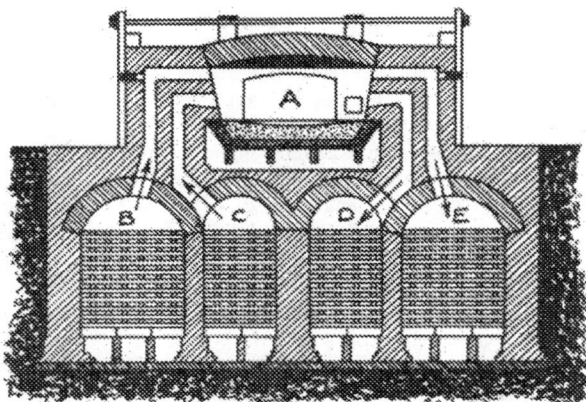

Fig. 118.—Siemens Regenerative Furnace: Diagram.

chamber and heated directly by the hot gases from the grate, which are deflected upon it by the arched roof. By regulating the supply of air an oxidizing or reducing action can be obtained at will. To obtain the former effect the firebars must be set well apart and the fuel fed in a thin layer, and for the latter effect the firebars must be set closer and the fuel fed in so as to form a thick layer.

The Regenerative Furnace.—Fig. 118 is a diagrammatic illustration of this type of furnace, which owes its in-

ception to Siemens. The object of this furnace is to
recover as much heat as possible from the flue gases.
To effect this the furnace is connected with a number
of chambers or flues filled with firebrick, through a
certain number of which the hot waste gases pass, thus
giving up their heat to the firebrick packing. After
about twenty minutes to half an hour the waste gases are
diverted by means of dampers to a fresh set of cool flues,
and at the same time the incoming gas and air is made
to pass through the heated flues and recover the heat
therein. In the glass-making industry the pot furnaces
are frequently of this type, although the recuperative
furnace, in which there is no reversal of draught, but
the incoming gas is made to pass over fireclay tubes
heated by the waste gases, is also in use.

Roasting Furnaces.—The chemical industry of this
country requires enormous quantities of sulphuric acid,
the production of which depends upon the oxidation of
huge quantities of sulphur. A great proportion of this
sulphur is obtained by heating ores which contain sulphur,
in specially constructed furnaces, which aim at producing
sulphur dioxide gas in as pure a state as possible.

Fig. 119 is an illustration of a mechanical roasting
furnace for copper and iron pyrites, spent oxide, gold
ores, silver lead ores, concentrates, zinc ores, etc., made
by the Harris Furnace Co., Ltd., Sheffield.

The furnace is built in vertical sections separated by
division walls, and each section is divided into the desired
number of tiers by arched floors. In each section there
are two vertical rabble shafts mounted on substantial
ball-bearing pedestals, which are adjustable for height
and separated from the interior of the furnace by an
arched opening accessible from the outside at any time.
There are two types of shafts used, known as the " A "
and " B " types, illustrations of which are shown in
Fig. 120 and Fig. 121 respectively. The " A " type is
so constructed that it can be easily taken to pieces, or any

one arm can be replaced without interfering with any other part of the shaft, by simply removing the bolts

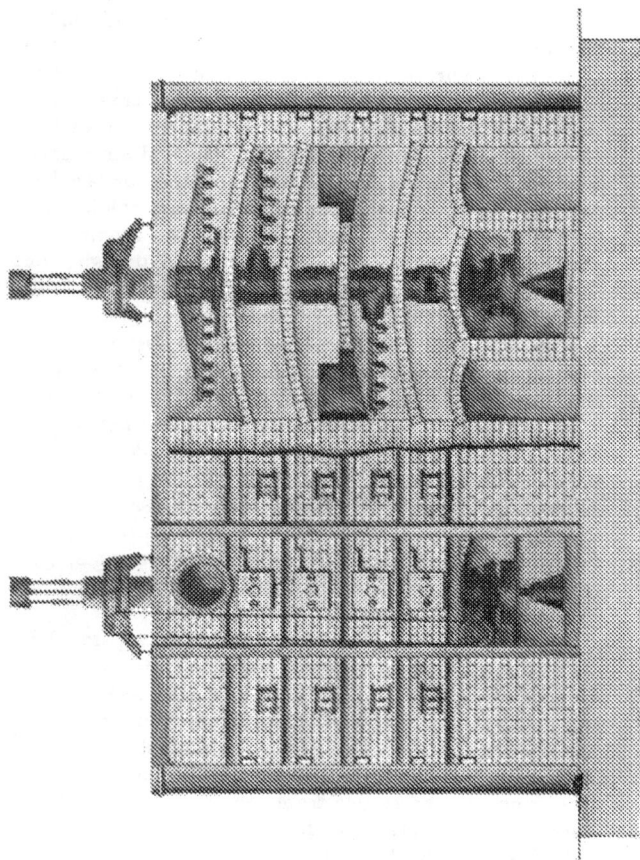

FIG. 110.—HARRIS MECHANICAL ROASTING FURNACE.

in the top and bottom joints. The shaft has a separate flow of water to each arm, and a return to the centre of the

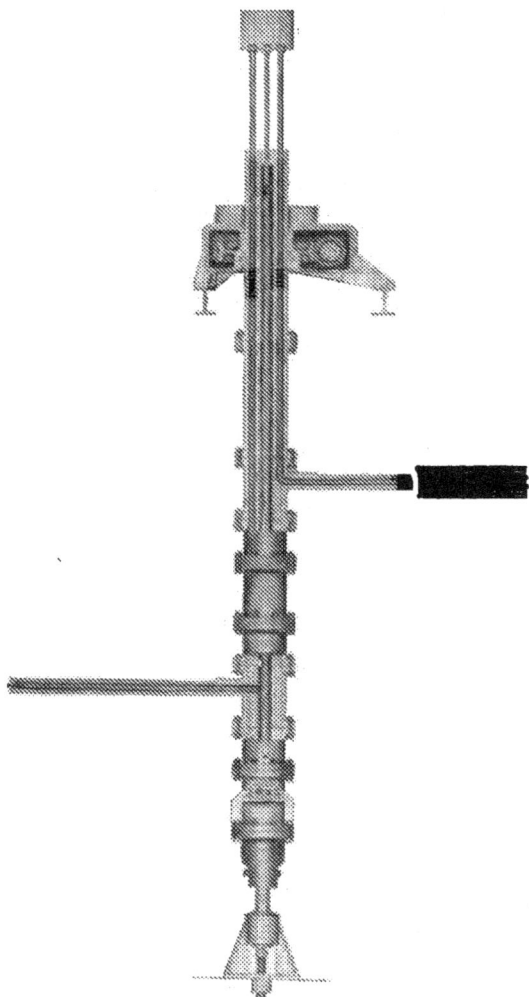

Fig. 120.—" A " Type Shaft for Roasting Furnace.

shaft, through which water is carried up, whence it is taken off at a lower level than the feed into the pan, and is then conveyed through suitable piping, to be disposed of as desired.

When only hard water is available for cooling, type " B " shaft is used, as the " A " type is liable to become choked with lime deposit, which, stopping the flow of cooling water, allows the arm to become red hot and possibly be burnt off, thus necessitating the closing down of the whole section. In the " B " type the arm is detachable from the shaft by removing the first rake in the arm, which rake also acts as a locking piece to the cover plate on the front of the boss. The cover plate not only holds the arm in position, but also prevents any gases from the furnace entering the shaft, or air in the shaft reaching the furnace. The arm can be either air or water cooled on any or all of the hearths, and in the latter case the water pipes are lowered into the arms from the top of the shaft, the joint thus being inside the shaft and obviating the possibility of water getting inside the furnace. The very simple construction on the top of the shaft is so arranged that no arm but the one to be operated on need be interfered with, the water pipes being lifted from the arm projection inside the shaft. The first rake and cover plate having been removed (working from the furnace door), the arm is then free of the shaft and can be drawn out and replaced, and the necessary repairs to the defective arm carried out as desired. Each arm has at least a $1\frac{1}{2}$-inch water way and a separate flow and return governed by valves at the top of the shaft, so that the heat from each arm can be tested and the growth of deposit observed. The water is taken off at the bottom of the shaft, thereby causing a current of cold air to travel up the shaft, and also relieving the arms of any pressure. When air-cooling is used the air enters the shaft at the bottom, and, passing through an opening in the bottom of the arm, it travels

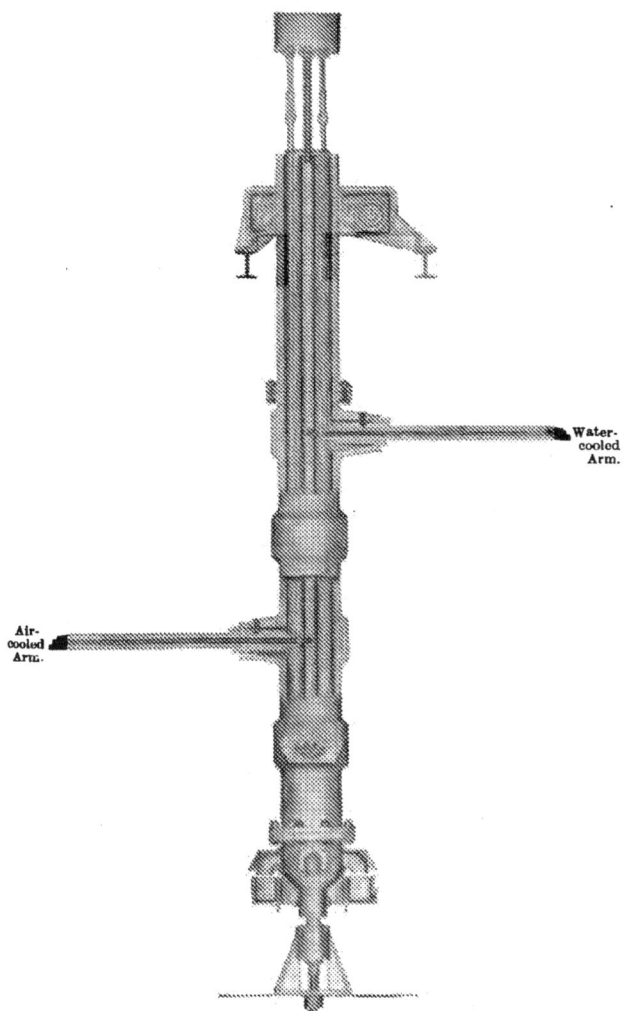

Fig. 121.—" B " Type Shaft for Roasting Furnace.

along and returns overhead; thence it re-enters the shaft from the top side of the arm and travels upwards, finding its exit at the top of the shaft. The illustration shows the lower arm being air-cooled and the upper arm water-cooled.

The rakes are of the slip-on type, and can be easily changed in a few minutes; consequently the pattern or pitch of the rakes can be so arranged that different depths of material can be maintained on each bed without interference with the discharge. Thus on the top bed, where the combustion is most rapid, a shallow working load can be maintained, while on the lower beds, where the sulphur is partly burnt off and it becomes necessary to retain all the heat possible, a deeper load can be kept with advantage. This interchangeability of the rakes is a great advantage in the roasting of spent oxide and different grades of copper or iron pyrites.

The ore to be roasted is fed through a suitable feeding arrangement in the roof of the furnace, adjacent to the centre of one of the shafts in the uppermost tier. The rakes on the first arm are so arranged that the ore is gradually moved towards the circumference of the arm path, whence it comes under the control of the other arm in that tier, the rakes on which are arranged to move the ore towards the centre of the arm path, whence the ore passes through a feed opening to the next tier. This operation is repeated in each tier until the ore is finally delivered from the lowermost tier into a discharge spiral conveyor or other suitable arrangement for dealing with burnt ore. The gas apertures are arranged at alternate ends of the hearths, and rakes are provided on the roof of the furnace for utilizing waste heat for drying damp ore or other material. In the event of repairs or renewals being necessary, the section affected is cooled by stopping the feed to the same and opening fully all its air doors, without interfering with the work of the remaining sections.

Driving belts are dispensed with throughout the whole furnace, and the separate sections are driven by claw-

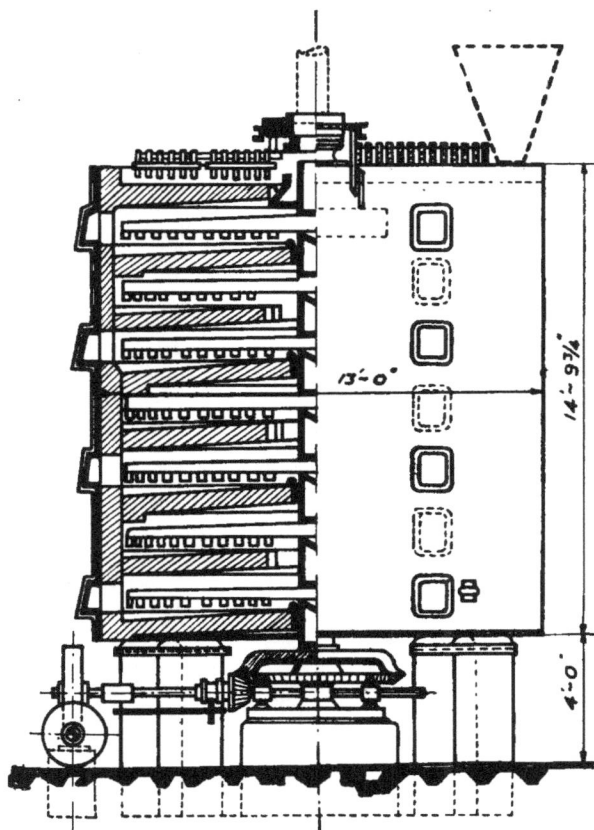

FIG. 122.—H.H. TYPE MECHANICAL ROASTING FURNACE.

clutch gears from a main shaft, which is in turn driven from the engine, motor, or existing line shafting. The

12

sections which are independent of one another require about 1 b.h.p. per vertical shaft, and the additional heat obtained in the Glover tower through having no separate dust chamber enables the whole make of the plant to be concentrated in the tower into acid of from 145° to 150° (T.).

The following are the capacities for a twenty-four hours' roast of various sizes of the Harris furnace:

Ground Space.	Copper Pyrites or Spent Oxide.
44 feet × 20 feet.	28 tons.
33 ,, × 20 ,,	21 ,,
23 ,, × 20 ,,	14 ,,
12 ,, × 20 ,,	7 ,,
	Australian Zinc Blende.
22 feet 6 inches × 22 feet.	12 to 14 tons.

One man can easily attend to two or three furnaces, roasting from 45 to 60 tons of ore per twenty-four hours.

Fig. 122 illustrates the type of furnace made by Huntington, Heberlein and Co., Ltd., London. It has a capacity of 5 to 5½ tons of 48 per cent. pyrites in twenty-four hours, according to the composition of the ore, and has an air-cooled shaft with natural draught and a top drying shelf. The furnace has a diameter of 13 feet, requiring 370 square feet of floor space, and has seven hearths giving a total hearth area of 624 square feet. The power required is ¾ h.p. and a dust-proof discharge and funnel are also provided when necessary.

CHAPTER VI

WATER TREATMENT PLANT

THE attempt to obtain a universal solvent engaged the major portion of the time of a great many alchemists. Had they been content with a comparatively slow action and with dilute solutions, they would have found that water was the nearest approach to their ideal that it was possible to find. This solvent property of water, together with the operation of the laws of mass action, should always be present in the mind of the chemical engineer. All natural waters are more or less impure, and the nature and extent of the purification required depends upon the uses to which they are put, which may be roughly classified as follows: (1) Food purposes; (2) the manufacture of industrial products; and (3) steam raising.

The method of removing insoluble material and matter held in suspension has already been dealt with in considering filtering apparatus, so that the matter of concern at the moment is the removal of those dissolved substances by methods other than those of distillation, also previously mentioned.

The presence of dissolved minerals in natural waters is the cause in boiler-room practice of the trouble of scale formation, corrosion, and foaming.

From the nature of the substances concerned, waters containing a certain amount of inorganic impurities are termed hard, and the process resorted to for their purification is called water-softening.

When hard water is evaporated the mineral impurities dissolved in it are precipitated, and settle upon the shell and tubes of boilers as hard scale, the rate of incrustation, its composition, hardness, and density, depending upon the quality of the water, the steam pressure, and other circumstances.

If we take the steam requirement of the average engine as being equivalent to 2 gallons of water per horse-power indicated per hour, and that the water contains 15 grains of scale-forming salts per gallon, which is less than is commonly the case, the scale deposited in a working day of ten hours amounts to about ¾ ounce per i.h.p.

Calcium and magnesium in the form of carbonates and sulphates form about 90 per cent. of the scale commonly found in boilers, which forms an insulating medium with a high power of resistance to heat. Professor Rankine estimates that the heat resistance of carbonate of calcium is seventeen times that of iron, and of sulphate of calcium forty-eight times that of iron. He therefore calculates that ⅛ inch of average scale necessitates the expenditure of 16 per cent., ¼ inch of 50 per cent., and ½ inch of 150 per cent., extra fuel to generate the same amount of steam, as compared with a clean boiler. It has been ascertained in this connection that, whereas the temperature of a clean boiler plate is only 350° F., the temperature of the same plate covered with ½ inch of scale is 750° F.—*i.e.*, 400° F. above the temperature actually required to convert the water into steam, involving the danger of collapse of the furnace crowns.

Corrosion or pitting is mainly caused by the presence of free acids in the original water or formed by the interaction of the solutes under certain conditions of temperature and pressure obtained in the boiler. Chlorides of the metals are particularly ready to dissociate and form hydrochloric acid in the presence of moisture, and the results of mass action become apparent wherever different phases of iron of the boiler are in contact, such as at the rivets. The most abundant chloride found in water is that of magnesium, and it is the frequent cause of serious trouble in boilers. Nitrates are also found, and also exert a similar corrosive action.

Foaming is essentially the formation of large masses of bubbles on the surface of the water in the boiler and in the steam space above, which do not break readily and release the steam. The strength of the film is dependent upon the nature of the water in the boiler, the steam pressure, and other conditions present, but as a rough guide the tendency to foam is measured by the concentration of sodium and potassium salts in the water. It is obvious that since surface tension is so readily a variable quantity, the prevention of foaming largely depends upon the skill and experience of the operators.

The following mineral impurities are of common occurrence in water:

Calcium Carbonate.—In its pure state it is only slightly soluble in water. It, however, dissolves freely in water containing carbonic acid, forming calcium bicarbonate. When water containing calcium bicarbonate is heated, the carbonic acid is driven off and the normal carbonate is precipitated. Calcium carbonate by itself forms a comparatively soft scale, but with other ingredients in the water forms a hard scale.

Calcium Sulphate.—This forms a hard flinty scale, and attaches itself very firmly to the boilers.

Calcium Chloride.—This substance is very soluble in water, and will not cause incrustation or deposit, but, being a chloride, it will readily react to form calcium sulphate, and also cause corrosion.

Calcium Nitrate.—This has a similar action to the chloride, and readily forms the sulphate, and also causes corrosion.

Magnesium Carbonate.—Has a similar action to calcium carbonate, its normal carbonate being sparingly soluble, while its bicarbonate is much more soluble.

Magnesium Sulphate.—It is very soluble in water and does not form a scale, but in the presence of calcium carbonate both calcium sulphate and magnesium carbonate are formed as scale.

Magnesium Chloride.—This does not form a scale, but, being a chloride, it readily forms hydrochloric acid, which causes corrosion.

Sodium Sulphate.—Is a very soluble alkaline salt which does not form a scale, but increases the tendency to foaming.

Sodium Chloride.—It behaves similarly to the sulphate, and is fairly stable at boiler temperatures, but, being a chloride, must be reckoned with accordingly.

Iron.—This is usually present in the form of the bicarbonate, which readily gives up its carbon dioxide and is oxidized to the hydroxide, forming a gelatinous scum. In acid water the sulphate may be present, but it is very readily treated.

Alumina.—Is found in small quantities in most waters.

Silica.—Is found in nearly all waters, and when present in appreciable quantities it unites with other ingredients to form an extremely hard scale.

Carbon Dioxide.—This is found in all natural waters in excess of that required to form the bicarbonates found in solution, and is the cause of a certain amount of corrosion.

Hardness which is caused by the presence of such substances as the bicarbonates which are precipitated on boiling is known as temporary hardness, the other salts producing permanent hardness, the two together making up the total hardness of the water.

Messrs. Sofnol, Ltd., Greenwich, who are experts in water-softening, very truly remark that softening is not so much a mechanical as a chemical process, and too much attention is generally paid to the mechanical part, whilst the chemistry is allowed to take its chance. The function of the machine is to bring the water into intimate contact with the proper amount of the chemicals, to remove the precipitates formed, and to deliver a clear effluent.

The chemistry of the process is the formation of new combinations, which, being insoluble, allow the machine to perform the mechanical part and deliver a clear, softened effluent.

The chemicals must be—

1. In a fine state of division.
2. As light as possible.
3. Quickly soluble.
4. In proper proportions.
5. Uniform in composition.

Further, they must act immediately, do their work as quickly as possible, and yield precipitates which readily settle.

If these conditions are fulfilled and the machine brings the chemicals into intimate contact with the water, then the water will be properly softened; but if these conditions are not fulfilled the process is a haphazard one, and the results are neither concordant nor satisfactory.

The process of softening a carbonate water is essentially different from that required by a sulphate water. In the first case the withdrawal of the free carbonic acid removes the solvent of the carbonates; they thus become insoluble and the water loses its hardness. On the other hand, the sulphates, being dissolved by the water itself, are not eliminated by the removal of the carbonic acid, and some other material requires to be added to cause them to become insoluble. Carbonate of soda is generally used for this purpose, but this cannot act so long as free carbonic acid remains in the water. Each grain of free carbonic acid means that every 1,000 gallons of the water will put 5 ounces of carbonate of soda out of action and prevent it doing its work as a destroyer of sulphate of lime and its analogues. Hence, unless the free acid is removed there is a great waste of soda, and the softened water has a high residual alkalinity, which will render it unfit for many purposes and manifest itself unpleasantly in the boilers.

Water-softening being a chemical operation, the factors of proportion, time, and temperature apply, and practical experience serves to show the many difficulties to be overcome. A great deal depends upon the selection of the proper type of plant for the work in hand, and when the water exceeds a moderate hardness the lime and soda type becomes a necessity. This type of plant, when properly designed and with due attention, will give satisfactory results.

There are two general types of lime-soda water-softening plants—the intermittent and the continuous types. The intermittent type, which possesses several advantages, usually consists of two large tanks, each tank holding at least four hours' supply of water. The process works intermittently, so that when one tank of water is being softened the other tank is being filled with hard water. The volume of water in the tank is known, and to this measured volume of water of known hardness a weighed quantity of lime and soda is added—the lime in the form of milk of lime and the soda in solution form. The contents of the tank are then thoroughly mixed and allowed to settle, when the calcium and magnesium compounds fall to the bottom of the tank. The clear softened water may then be drawn off for direct use or into a store tank, and the precipitated solids drawn off by an outlet in the bottom of the tank.

Owing to the facilities for control, this is probably the most exact method of water-softening, and, with an ordinary hard water, practically the whole of the hardness-forming salts can be removed, and the softened water contains the minimum excess of lime and soda. In a works where there is sufficient room for tanks, and first cost is not essential, it forms probably the most satisfactory plant.

The continuous-process plant, taking less room and requiring less attention, has been more developed. For successful working it must conform to the following conditions:

1. The lime and soda control system must work accurately.

2. The tank capacity must be large enough to enable the chemical reaction to complete itself fully.

3. The lime and magnesia sludge must be easily removable from the plant.

Fig. 123 gives a view of a rectangular form of the "Lassen-Hjort" automatic water-softener made by the United Water Softeners, Ltd., London.

This apparatus is designed to perform the following functions: `

1. Measurement and proportioning of the water.

2. Measurement and proportioning of the chemicals.

3. Settlement and filtration of the precipitate.

4. Regulation of the supply of both untreated and softened water.

The main parts of the apparatus are the mixing and measuring apparatus and the settling tanks and filters.

The measuring apparatus (Fig. 124) operates by leading the hard water into the plant by a pipe which alternately fills each of the compartments of a two-chambered tipper oscillating on a shaft carried in bearings. When one of these compartments is full of water the disturbance of equilibrium causes the tipper to overbalance, and, by doing so, to discharge its contents into the tank in which it is suspended. At the same time the other compart-ment of the tipper is brought under the orifice of the inlet pipe and filled in its turn with hard water, to be discharged in the same manner when full. As a definite quantity of water is passed at each oscillation, by at-taching a counter to the tipper shaft, the quantity of water passing through the plant can be accurately determined.

At each discharge of water from the tipper into the

FIG. 123.—RECTANGULAR WATER-SOFTENING APPARATUS.

tank a corresponding amount of water is displaced from this tank through a standpipe and shoot into the reaction chamber, and here it receives at the same moment the requisite charge of chemical solution from the semi-circular container affixed to the side of the tipper tank.

FIG. 124.—AUTOMATIC WEIGHING AND MEASURING APPARATUS FOR WATER-SOFTENING APPARATUS.

This is effected by the positive discharge valve placed in the bottom of the chemical container, which is opened at every movement of the tipper, and caused to deliver into the reaction chamber the exact amount of softening reagent (in the majority of cases a mixture of lime and soda ash) required to soften it to the guaranteed figure. The valve can be adjusted to deliver any specified quantity of reagent required by the volume of water in the tipper.

In the illustration of this valve (Fig. 125), *A* is a cylinder fixed to the bottom of the chemical reservoir, into which screws an adjustable cylinder *B*, secured in any desired position by the back nut *C*. Within these two cylinders work two valves, *D* and *E*, the latter screwing on to a tail piece *F*, projecting from the valve *D*. The pitch of the threads on this tail piece and the adjustable cylinder being the same, any movement of the cylinder *B* results in a corresponding movement of the valve *E*, owing to the valve *E* having a feather *G* working in a key-way *H* cut into the cylinder *B*. The valve *D* is provided with a flat face and a piston body, which latter prevents any chemical solution being admitted into the adjustable cylinder until the lower valve *E* has closed the outlet ports *J*. The operating gear consists of a double lever *K* fixed to the rocking shaft *L* of the tipper. These levers are fixed to the vertical valve spindle by two loose links *M* and trunnions *N*, clamped against a screwed sleeve *O* by the lock nut *P*. These levers, when in operation, impart an up-and-down motion to the valve. The screwed sleeve *O* works between rollers *Q* carried on to the bridge *R*. The object of the weight *S* is to keep the valve *D* tight on its seat.

The oscillating receiver is prevented from tipping until it contains a predetermined quantity of water by means of a locking gear constructed as follows: To the end plate of each compartment of the tipping bucket is attached a bracket carrying a ball float and lever, and a vertically sliding rod actuated by these, which latter at a certain height of the water lifts a lever

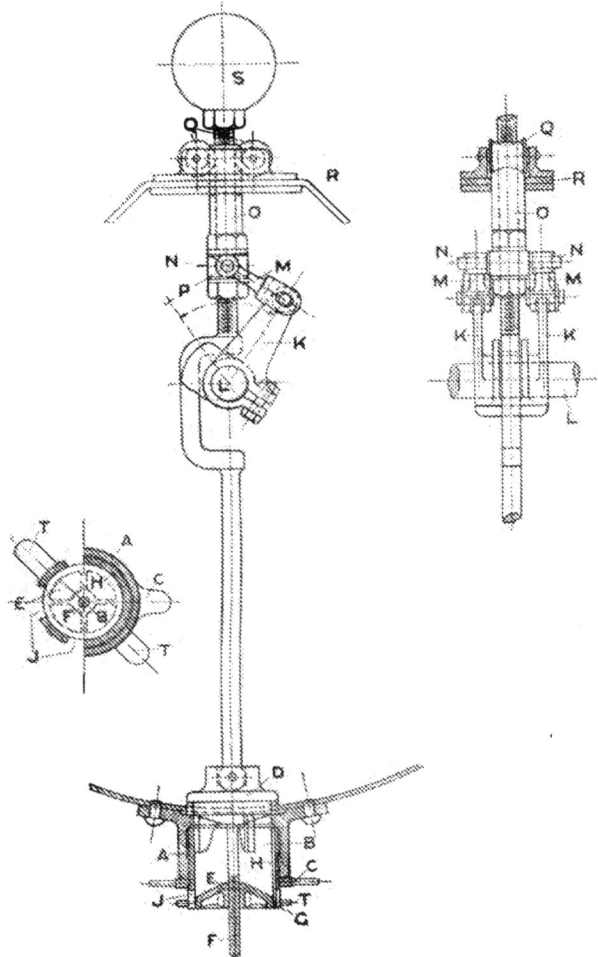

FIG. 125.—POSITIVE DISCHARGE VALVE FOR WATER-SOFTENING APPARATUS.

fulcrumed on the angle-iron edge of the tank, and engaged with a notch provided on the bracket before mentioned. On further rising, the lever is disengaged from the notch and the bucket tips. A link from the end of the tipper shaft operates a counter, which registers the number of tips.

The heavier portion of the precipitate produced by the addition of the measured quantities of softening reagents to the water settles to the bottom of the reaction chamber, whence it is removed daily by opening the sludge cocks. The finer precipitate, which will not settle, is retained in filters consisting of wood fibre packed between two rows of wood bars. The filters require cleaning, on an average, about every two months, which operation is effected by removing the top bars, by loosening the fibre, and washing it through with water.

The chemical container of the softener is designed, in the majority of cases, to hold eight to twelve hours' supply, when it can be refilled from a mixing tank situated in any convenient position.

Fig. 126 shows a cylindrical type of softener which is useful where soft water is required to be delivered at a height. The operations involved are the same as previously described, but the water from the measuring apparatus passes down a central tube, depositing precipitate as it slowly rises up the tank and through the filters to the storage tanks.

Permutit.—This is the name given to an artificial zeolite having the formula $Al_2O_3.10SiO_2.10Na_2O$, and marketed by the United Water Softeners, Ltd., London.

The valuable property of this substance is the readiness with which it will exchange its sodium for calcium and magnesium, the reaction being reversible, and therefore solely a question of mass action. Therefore, if hard water is passed through a bed of Permutit, a calcium-magnesium Permutit is formed, and only sodium salts pass through;

^{but} although the hardness is removed, the total amount
^{of} solids remains the same.

When the sodium of the Permutit is exhausted by the

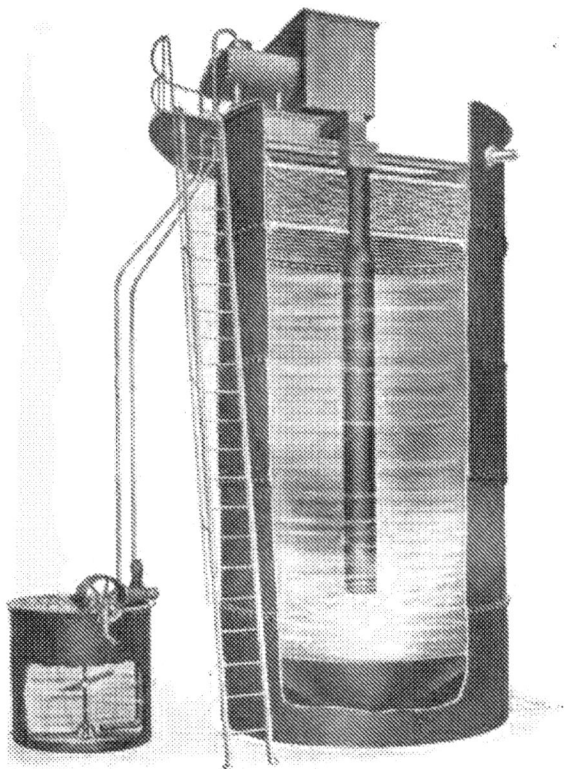

FIG. 126.—CYLINDRICAL WATER-SOFTENING APPARATUS.

replacement with calcium and magnesium, it is treated
with a solution of salt, which by mass action converts
the Permutit back to its original condition.

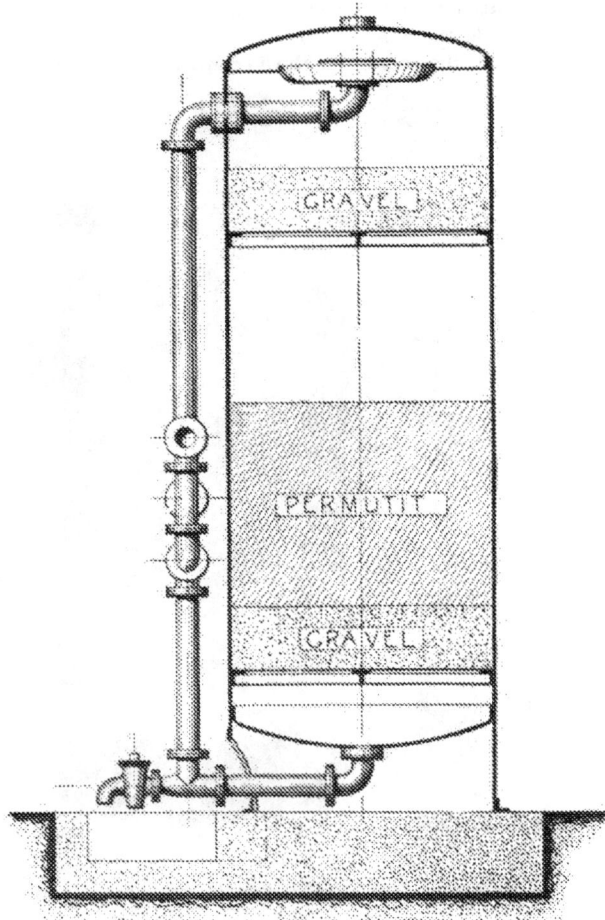

Fig. 127.—Permutit Water-softening Apparatus: Diagram.

The simplicity of the chemical reaction finds its counterpart in the extreme compactness and convenience of the apparatus requisite for the softening process. A Permutit softener, of which a sectional view is given in Fig. 127, consists simply of a cylinder to contain the Permutit, connected with a receptacle holding the salt solution for regeneration, and fitted with the necessary valves for controlling the water flow.

The design of the plant and the nature of the Permutit process favour the carrying out of the softening under the ordinary pressure of the water mains, thus giving this system the great advantage that it can be connected to the water main in any position without the necessity of pumping twice, or of arranging for a gravity flow of softened water.

The removal of iron from water is accomplished by a Permutit in which the sodium is replaced by an oxidized product of manganese which oxidizes the iron to the hydrate which is retained by the filter. When the oxidizing properties of the Permutit are exhausted they are restored by means of a solution of potassium permanganate.

The Permutit process will give water of zero hardness, which is of inestimable advantage in many industries, such as silk and cotton dyeing and bleaching, wool scouring, laundry work, etc.

For food and drinking purposes water is sterilized by injecting liquid chlorine and removing the excess by sulphur dioxide, and the plant is mostly of the nature with which the chemical student is already familiar.

13

CHAPTER VII

THE CONTROL OF TEMPERATURE

In most chemical industries reactions have to be carried out at definite temperatures, and as these temperatures are in most cases well above the normal atmospheric temperature, some means of temperature control is necessary, so that the process may be worked efficiently with the least possible consumption of heat units. The commonest form of heating is by means of steam, and James Baldwin and Co., Keighley, have devised a system of temperature control which, applied to steam, acts independently of the boiler and controls the steam supply during boiling and other processes. This device is known as the "Isothermal" (electric-mercury thermometer-control) valve, which is claimed to regulate automatically temperature to within 1° F.

In the case of steam the supply is controlled by electricity, so that the parts of the control, consisting of the valve, thermometer, and the transformer, may be separately fixed in suitable positions.

The control valve, of which a section is shown in Fig. 128, is fixed in a horizontal position, on the steam pipe on the outlet side of the usual steam stop valve, the internal parts operating vertically, so that when not in use the valve is free from pressure. As illustrated in the section the valve is in the normal position, closed, there being no pressure. Instantly steam is admitted the action of the pressure upon piston C lifts the piston valves D D full open, allowing a full-bore passage for the steam at any or varying pressures. The valves D D are of the equilibrium type, and are cast on the same stem

as the pistons B and C, the piston C being smaller in
diameter than the piston B.   In the cover of the valve-

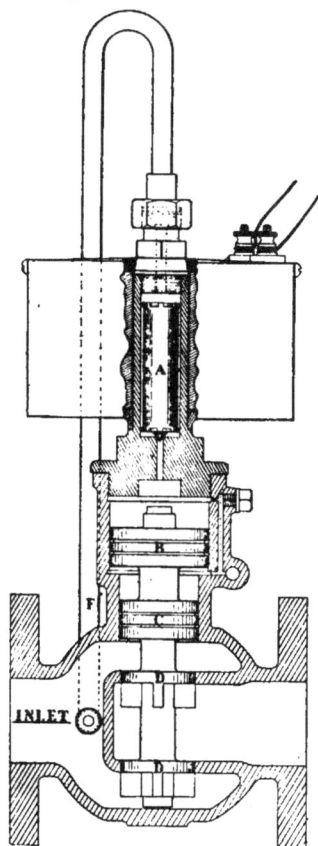

Fig. 128.—" Isothermal " Steam Valve.

body is a small valve A, which is in the centre of a
solenoid.   There is a connecting pipe F to the valve A

from the inlet or pressure side of the pistons C, $D\,D$. When pressure is admitted to the valve the pistons C, B are forced up against the cover or end of the cylinder which forms the outlet side of the small·.valve A, and the valves $D\,D$ are fully opened. If the valve A be

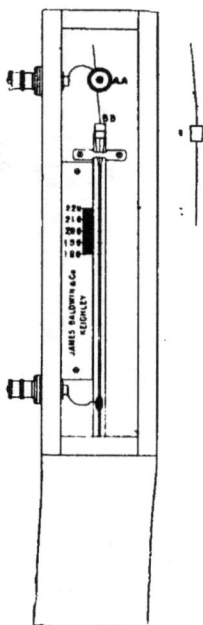

FIG. 129.—" ISOTHERMAL " THERMOMETER.

raised from its seating, the inlet pressure passes along the pipe F, and, acting on the large piston B, forces down into their seatings the valves $D\,D$, shutting off the steam supply. If the small valve A be allowed to fall on to its seating, the pressure upon the large piston B is removed (a small escape for this pressure is situated

between the valve A and piston B), and the boiler pressure
acting on the piston C opens the valves $D\,D$ full bore.
It will be noted that the pistons B, C and the valves $D\,D$
are operated by steam pressure, while the valve A in the
solenoid is operated by electricity.

A mercury-column thermometer acts as an electric
switch, and is specially constructed, having terminals
for connections, as shown in Fig. 129.   The thermometer,
suitably calibrated, is fixed in a selected position on the
vessel of which the temperature of the contents is to be

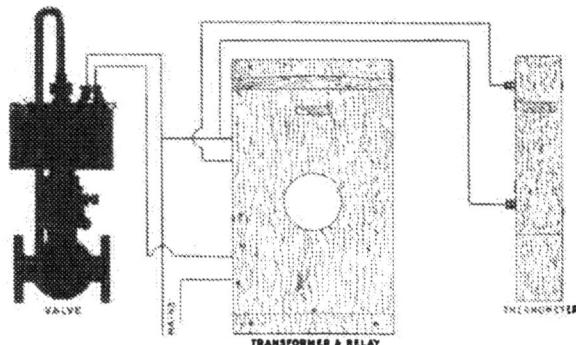

FIG. 130.—G.A. " ISOTHERMAL " TEMPERATURE CONTROL
APPARATUS.

controlled.   The bulb of the thermometer is connected
to one terminal, and the top is closed by a cork carrying
a platinum wire which can be adjusted so as to make
contact with the mercury at any desired temperature.

This switch operates the solenoid in the valve through
a transformer and relay connected up as shown in Fig. 130,
the current required for each valve being about 0·6
ampère at 110 volts, and for the thermometer 15 milli-
ampères at 1 volt.

In case of failure of the current supply the valve is
free to open, and the steam supply can be manipulated by

hand in the ordinary manner, or the valve can be constructed to shut off automatically the steam supply
until the current is again restored.

The very wide range of application of this system of
temperature control includes the steam heating of factories.

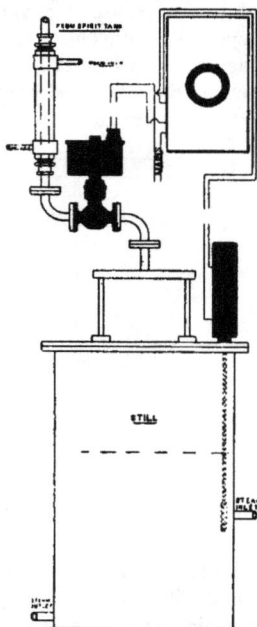

FIG. 131.—" ISOTHERMAL " CONTROL OF STILL.

etc., hot water heating, and the cooling of vessels, by passing
cold water to maintain a lower temperature required in
many chemical processes. Fig. 131 is a diagram showing
its application to a still, and Fig. 132 its application
to a jacketed pan. An exceedingly important application
is its use in a dye vessel for dyeing piece goods, to which

it is connected as shown in Fig. 133. It can also work as a reducing valve for maintaining an exhaust steam pressure at a constant pressure of, say, 5 pounds per square inch, by admission of high-pressure live steam as shown in Fig. 134. Applied to a vulcanizing press, pan, or cylinder, as shown in Fig. 135, it will maintain a constant temperature and reduced steam pressure, or, as shown in Fig. 136, maintain the blast to gas producers at

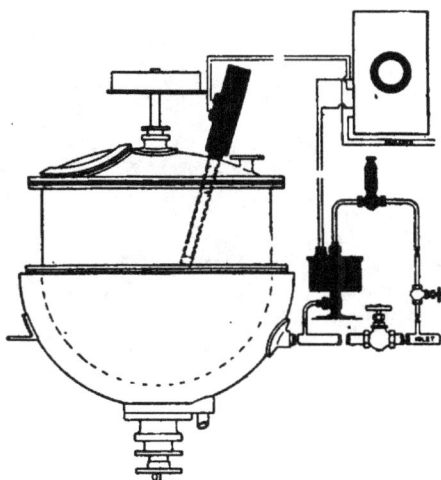

FIG. 132.—" ISOTHERMAL " CONTROL OF STEAM-JACKETED PAN.

constant temperature and composition. Fig. 137 shows how this apparatus can be used for the automatic regulation of the temperature and humidity of cotton-spinning and other rooms. Fig. 138 is a diagram of the valve used for regulating the temperature of superheated steam by the admission of saturated steam, and Figs. 139 and 140 illustrate in section the valve for regulating the gas supply for sizes $\frac{3}{32}$ inch to $\frac{1}{2}$ inch bore and $\frac{3}{4}$ inch to

FIG. 133.—" ISOTHERMAL " CONTROL OF DYE VESSEL.

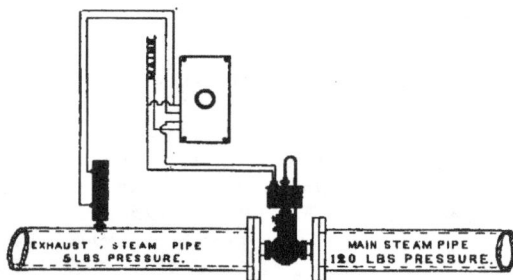

FIG. 134.—" ISOTHERMAL " CONTROL OF EXHAUST STEAM.

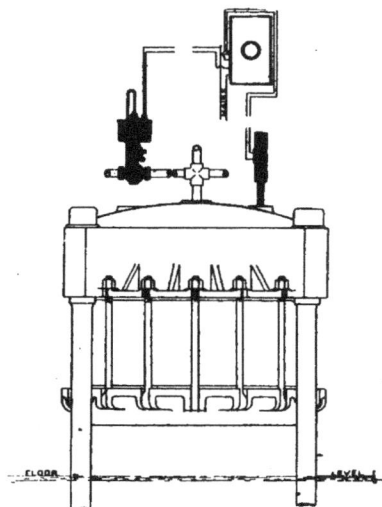

FIG. 135.—" ISOTHERMAL " CONTROL OF VULCANIZING PAN.

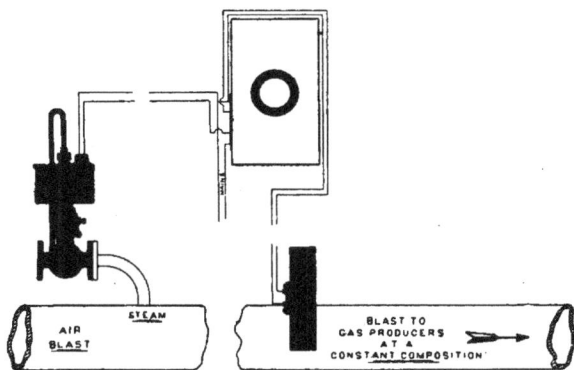

FIG. 136.—." ISOTHERMAL " CONTROL OF BLAST FOR GAS PRODUCER.

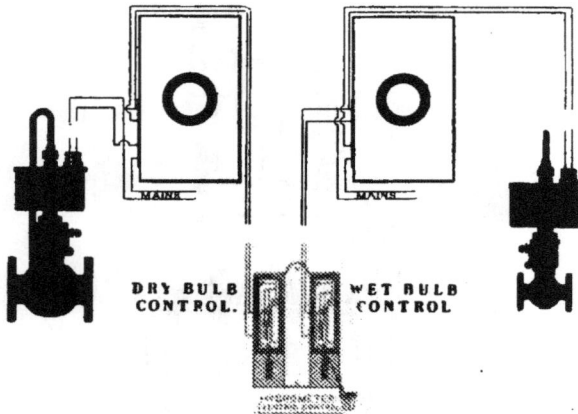

FIG. 137.—" ISOTHERMAL " CONTROL OF COTTON-SPINNING ROOMS.

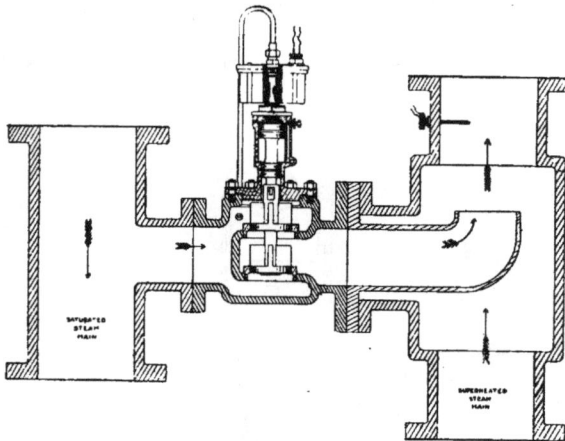

FIG. 138.—" ISOTHERMAL " SUPERHEATED STEAM VALVE.

2 inch bore respectively for gas-heating systems, and automatically regulating the temperatures up to 600° F. It may be used to regulate a fan so that it operates at any desired temperature, or as a steam main isolating valve, or for winding engines and for other purposes too numerous to mention.

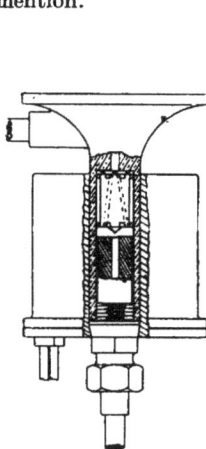

FIG. 139.—" ISOTHERMAL " GAS VALVE.   FIG. 140.—" ISOTHERMAL " GAS VALVE.

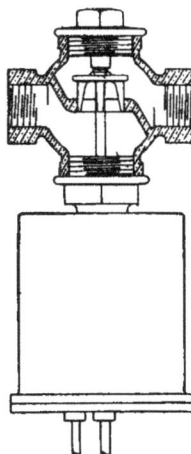

Refrigerating Machines.—The problem of maintaining temperatures below the normal atmospheric temperature demands the use of refrigerating machinery such as is made by the Lightfoot Refrigeration Co., Ltd., London.

The cold storage of food is familiar to all, but this forms only a portion of the field of application of the mechanical production of cold.   In dyeworks refrigerating machinery is indispensable for producing fast colours and even for colouring with certain dyes.   The supply of natural silk is considerably augmented by its use to regulate the hatching out of the eggs to suit the supply of food available for the silkworms, while artificial silk also depends upon it in the process of converting pulp into silky fibre. It forms an essential part of the equipment of the bacon-curing factory, the brewery, and the margarine factory, to name but a few.

In the Lightfoot system of refrigeration cold is produced by the evaporation of liquid ammonia or carbon dioxide, the vapour formed being afterwards condensed and used over again. Fig. 141 is a diagram illustrating the principal parts of which each machine consists.

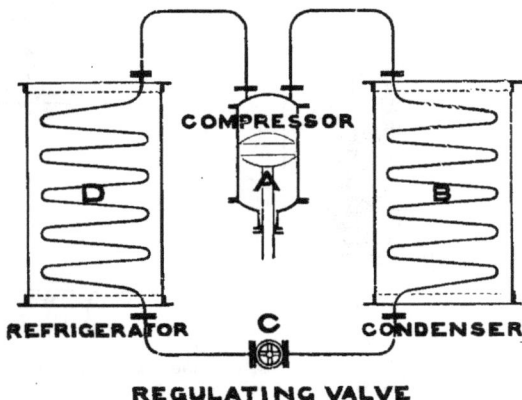

FIG. 141.—DIAGRAM OF LIGHTFOOT REFRIGERATION SYSTEM.

The refrigerator consists of a series of coils of special welded tube wound each in one length so as to avoid inaccessible joints, inside which the liquid ammonia or carbon dioxide, entering through a regulating valve, is vaporized, thus reducing the temperature of the liquid or material surrounding the coils.

The condenser, of which a normal open type is shown in Fig. 142, consists of a series of coils of special welded tubes, inside which the compressed vapours are cooled and liquefied, the liquid being returned to the refrigerator through the regulating valve. In order to cool the vapours the condenser coils are either completely immersed in water contained in a wrought-iron tank, or a spray of water is caused to trickle over the surface of the coils. The ammonia compressor, of which a small horizontal type is shown in Fig. 143, consists of a cylinder of tough, close-grained cast iron, with back and front covers of the same material. These covers contain the suction and delivery valves, which are turned out of solid steel, each fitted into a box in which is formed the seat,

FIG. 142.—OPEN CONDENSER.

the arrangements being such that any valve can be readily withdrawn and replaced without disturbing the connections. The piston rod is of polished steel, secured to the piston, and arranged to work through a special stuffing box formed in the front cover. The cross-head is of wrought iron, provided with a slipper with large wearing surface fitted with bronze bearings, and the connecting rod of polished wrought iron is fitted with white metal bearing at the large end. The bed plate is of cast iron, upon which are formed the guides for the cross-head and the bearing for the crank-shaft, which is ring lubricated. Special oiling arrangements are provided for lubricating the piston and for preventing the oil from passing over into the coils of the condenser and refrigerator.

The carbonic acid compressor, of which a vertical type is shown in Fig. 144, has a cylinder machined out of a billet of solid steel, as also are the covers and all the fittings. In the delivery valve cap is placed a safety valve, which will relieve the pressure in the event of the machine being started up with the delivery stop valve shut, but which will only allow just sufficient gas to escape to keep the pressure within safe limits. Cup leathers, which are a frequent source of trouble, are avoided by having a metallic packed piston and gland. In some cases the cast-iron frame forms a casing in which are contained the condenser coils, thus making a compact machine.

Ice-making.—The appearance of the ice produced is dependent on the water used and on the method of freezing—e.g., ordinary fresh water frozen without agitation produces opaque ice.

There are three systems on which ice is made viz., (1) the can ice system; (2) the cell ice system; and (3) the plate ice system.

The can ice system employs a number of cans of lead-coated steel, of rectangular section, tapering slightly from top to bottom, in which the water to be frozen is placed. These cans are placed in a tank containing brine sufficient to immerse them to within 2 or 3 inches of their tops. The brine is maintained at a temperature of 16° to 20° F. by means of the refrigerator coils of the machine, and circulated round the cans until all the water in the latter is frozen. The cans are then removed in rows and dipped in warm water for a short time to loosen the ice, so that it may be tipped out.

Fig. 143.—Horizontal Ammonia Compressor.

Fig. 144.—Vertical Carbon Dioxide Compressor.

The can system of ice-making is the least expensive in first cost, and also the most economical to work, and the ice formed is either opaque or, by the adoption of agitation gear, may be rendered clear to within about 5 per cent.

In the cell system a wooden tank is used, in which are placed a number of galvanized iron cells from 10 to 14 inches apart, through which cold brine is circulated. The space between the series of cells is filled with water which, during freezing, is made to oscillate gently up and down. Ice forms on the sides of the cells, gradually increasing in thickness until the two plates of ice on opposite cells join in the centre to make a single block. Warm brine is then circulated through the cells to loosen the blocks, which are lifted out of the tank by means of hooks or ropes frozen into them. Cell ice is made in blocks weighing from 4 to 6 cwt., and from 9 to 14 inches thick, which are of convenient size for handling.

The plate ice system consists of placing flat hollow walls of galvanized sheet iron in a large wooden tank which is filled with the water to be frozen. Brine or ammonia is circulated through the hollow walls, causing a plate of ice to form on each side of them, the water being agitated by means of compressed air. When the plates of ice are of the desired thickness—say, 12 or 14 inches—warm brine or ammonia is pumped through the walls, so as to loosen the ice and permit of its being withdrawn from the tank. Plate ice is transparent, and is the finest quality obtainable.

Cold Storage.—There are three systems in use, known as—(1) The brine pipe system; (2) the direct expansion pipe system; and (3) the air circulation system.

The brine pipe system may be used with advantage for certain purposes, such as for cooling, fermenting and storage cellars in breweries, bacon-curing beds, etc. The pipes are placed under the ceilings and sometimes on the walls of the rooms to be cooled, and cold brine from a brine refrigerator is pumped through them. An advantage of the brine system is that the large volume of cold brine in the pipes will maintain the low temperature in the rooms for a considerable time after the refrigerating machine has stopped.

The direct expansion pipe system consists in placing the ammonia or carbonic acid refrigerator coils directly in the cold rooms, these coils being arranged in a similar manner to the pipes in the brine system. The refrigerating agent is vaporized in these coils, thereby reducing the temperature of the chamber.

The air circulation system consists in circulating a current of pure, cold dry air at the desired temperature through the cold rooms. There is entire absence of snow, moisture, or drip in the rooms, and they are kept dry and free from smell. The apparatus is simple and less costly than the brine system, and owing to its compact arrangement much larger cooling surfaces are obtainable, with a consequently much increased efficiency. The apparatus being external to the rooms, its full power can be applied to any room without loss of efficiency, whereas with the pipe system the pipes in the rooms that are not being cooled are useless.

Lubrication is one of the practical difficulties of refrigerating machinery, and great care should be exercised in the choice of a suitable lubricant. The condenser coils of every machine should be examined at least once every year, and should the slightest sign of corrosion or pitting be discovered at any part of the coil, this part should be carefully cleaned and painted with two coats of some reliable bitumastic solution or protective paint.

The Absorption System.—In this system the cooling effect is also produced by the evaporation of liquid ammonia, but the cycle of operations is more extensive.

Aqueous ammonia is boiled by means of steam-heated coils in a still, and the vapours pass upwards through an analyzing column, where they meet a descending

14

stream of strong liquor which robs them of some of their moisture. The vapours pass thence to a dehydrator, where complete drying is effected, and the ammonia passes to the condenser to be liquefied and used for cooling purposes. The expanded gas then enters an absorber containing weak ammonia solution, which, when strengthened, flows into a tank, and is thence pumped to a heat exchanger, where it is raised to within 30° or 40° F. of the still temperature. It is then discharged down the analyzer into the still, and is again boiled and the cycle of operations repeated.

CHAPTER VIII

TRANSPORT

THE method adopted for the movement of material from one part of a chemical works to another depends upon the nature of the material, whether it is a solid or a fluid. Fluids possess inherent advantages for transportation, and it may be stated, in a general way, that the development of the methods of transporting solids has been along the lines of obtaining mechanical fluidity.

Conveying Solids.—The simplest, least efficient, and most common method adopted for the conveying of solids takes the form of the man-handled wheelbarrow. A barrow weighs from 60 to 70 pounds, and with this a man can move about $\frac{1}{2}$ ton of material 100 yards per hour over a level or slightly inclined surface. The proportion of shovellers to wheelers is determined by the nature of the particular job, but it may be taken that on an average a man can wheel a barrow having a capacity of 2 cubic feet about 200 feet a minute, and that it takes from one to two minutes to fill the barrow and about the same time to unload it. Barrows weighing up to 250 pounds, and having a capacity of 9 cubic feet, are sometimes used, and these are provided with two wheels as a rule. Steel plateways are frequently provided to ease the running and to save the track when much barrow work is done in any area.

Tipping Waggons.—For dealing with heavy loads a more efficient method is to use light four-wheeled cars provided with tyres suitable for running on steel rails and drawn by a small locomotive or electric motor. The

211

shape of the cars and the method of dumping vary considerably, but the commonest forms are the **V**-shaped

FIG. 145.—SIDE-TIPPING WAGGON.

cars for side tipping, as shown in Fig. 145, and end-tipping cars, of which a sample is shown in Fig. 146.

FIG. 146.—END-TIPPING WAGGON.

A cheap and efficient method is to run the cars on runways or overhead rails carrying a small trolley, from which the skip or bucket is suspended by means of a hook.   This method, which is illustrated in Fig. 147, has a very wide range of application.

Aerial Wire Ropeways.—During the last decade there has been a great development of this means of transport, and there is a constantly increasing field of employment for these ropeways, which are now designed with great efficiency in working, combined with largely reduced working costs, provided that care be exercised in adopting the best type of ropeway for any given duty.

Fig. 147.—Runway for Mine.

To give an account of this system even in outline would take up more space than is available at the moment.   It may be very inadequately described as consisting of a cable or wire rope, usually endless, which is suspended from towers, and along which are run carriages to which skips or buckets are attached.   Fig. 148 illustrates the sectional portable ropeway, for inter-works traffic, made by R. White and Sons, Widnes, who are specialists in aerial wire ropeways.

Among the various systems of aerial ropeways reference must be made to the single and the double rope system, the former in which the single rope acts both as the hauling and the carrying rope, and the latter where one rope does the hauling and the other the carrying of the load.

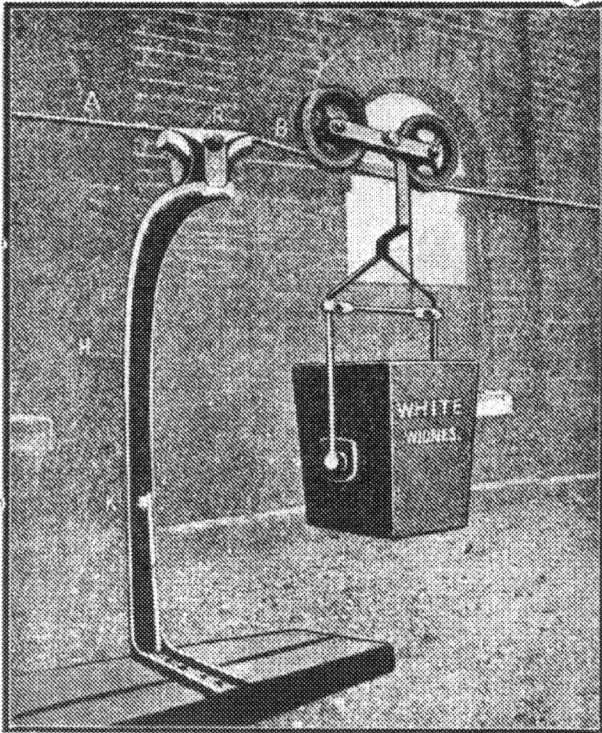

FIG. 148.—INTERWORKS TRAFFIC: PORTABLE ROPEWAY.

In the single-rope system an endless rope passes round a grooved driving wheel 6 to 12 feet in diameter at one end, and at the other end round a similar wheel kept up by a tension weight so that a constant tension is put on the rope. About every 100 yards the wire is supported by standards with cross-heads having four sheaves on the loaded side and two sheaves on the empty side. These sheaves are fixed to arms pivoted finally at the centre of the cross-head itself, so that each sheave as it receives the load is depressed and the weight of the rope is distributed over the other three sheaves. This arrangement can be clearly seen in the illustration of

FIG. 149.—STANDARD FOR SINGLE-ROPE SYSTEM.

one of the standards of a single-rope installation in Fig. 149. Cars are attached to the rope by means of a saddle, of which there are many types designed for quick engaging and disengaging and for maintaining a firm grip on inclines.

This type of ropeway is simple and efficient, and for straight lines with easy gradients and moderate loads is the best form, as it requires little attention and is practically fool-proof. The defects are that it cannot automatically negotiate horizontal angles, nor can the carriers be automatically carried round the return terminal, from the outwards to the return rope.

The double-rope system has been greatly developed to deal with severe gradients, heavy loads, automatic angles, automatic tipping, and automatic return of the carriers.

In this system two separate fixed ropes are used, stretched from one terminal to the other, and supported

FIG. 150.—STANDARD FOR DOUBLE-ROPE SYSTEM.

on standards, as shown in Fig. 150, placed usually from 100 to 200 yards apart. One end of each rope is generally fixed, whilst the other is led over guiding sheaves and terminates in a heavy tension weight. Saddles are fixed to the cross-heads about 6 to 12 feet apart, each saddle being long enough to support 2 to 3 feet of the rope, which lies in a groove. A fixed curved rail at each terminal connects the ends of the fixed ropes and completes the circuit. The carriers travel along these

ropes by means of two or more wheels fixed into the
head of the carrier, and a separate and lighter rope is
used to haul the carriers along. This haulage rope is
driven and kept tight as in the single rope system, and is
supported at the standards by rollers.

Elevators.—The bucket elevator is the machine
commonly used when it is desired to lift material any
distance. It consists of a number of buckets fastened
to an endless belt or link chain which passes over a wheel
in the hood at the top and another wheel in the boot
at the bottom. Power for driving is applied to the top
wheel in order to keep the loaded side taut, and the
whole machine is lightly boxed in with wood or sheet
metal in order to prevent the dispersal of dust. According to the nature of the material operated upon,
the buckets are made of steel, copper, or malleable iron,
and may be perforated to admit of drainage or have a
toothed edge to assist in raising fibrous materials. For
working with pasty material L-shaped buckets are
the best, as they are readily emptied, but the V-shaped
bucket is more commonly used, as it has a larger capacity;
in any case the buckets must be so arranged on the
chain or belt that at a given speed they discharge the
whole of their contents by centrifugal force as they go
over the top pulley. The material is discharged through
a spout in the hood, and the machine is capable of being
inclined so that the buckets give a clean discharge. When
the material to be lifted is in small particles and not of
a wearing nature, webbing or leather belts are used,
to which the sheet steel buckets are riveted. Such
machines are used in sizes having a bucket from 3 to 10
inches wide with an internal pulley from 9 to 24 inches
in diameter, driven at 90 to 40 revolutions per minute
When the material is in a rough condition a link chain,
such as is shown in Figs. 151 and 152, is used, and the
machine is driven at a slower speed than is the case when
webbing or leather bands can be used. Some form of

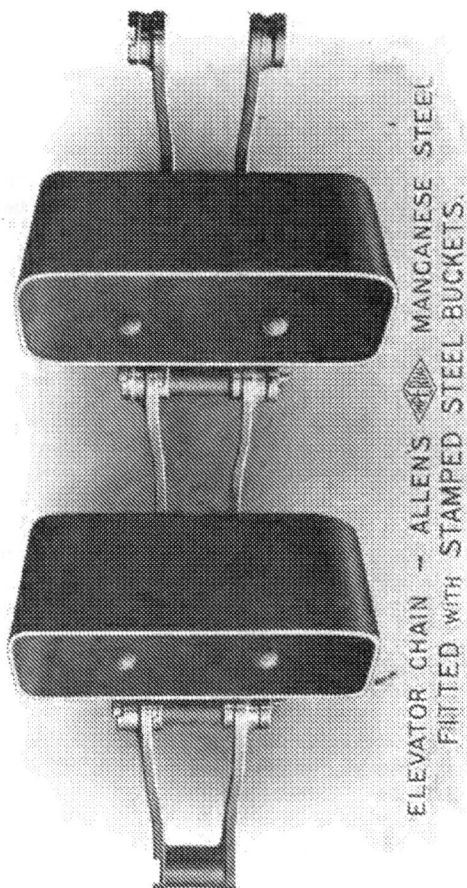

ELEVATOR CHAIN – ALLENS MANGANESE STEEL FITTED WITH STAMPED STEEL BUCKETS.

Fig. 151.—Elevator Chain.

tightening device is located at either end, according to whether interference with the feed or the drive is of lesser moment. The boot is usually constructed so that

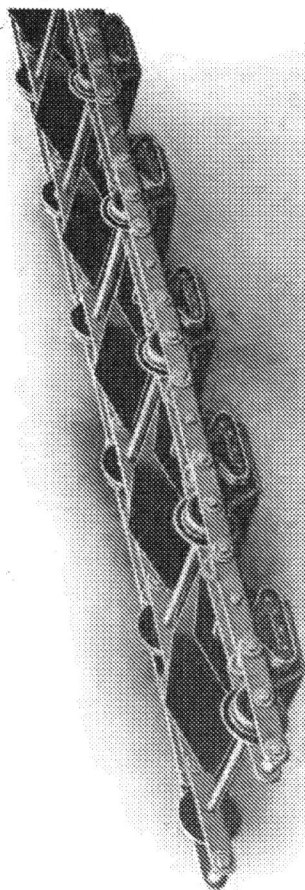

GRAVITY BUCKET CHAIN — ALLENS MANGANESE STEEL FITTED WITH GRAVITY BUCKETS.

Fig. 152.—Gravity Bucket Chain.

FIG. 153.—DUST-PROOF ELEVATOR.

the buckets can fill themselves by scraping up the
material as they travel round the bottom pulley, but it
is very desirable to give the buckets a direct and positive

feed when practicable. Owing to the slow speed, it
is found that with this type of elevator it is necessary
to give it a considerable inclination in order to get a
proper discharge, and for this reason the gravity type of
bucket is employed, so as to obtain a full load.

FIG. 154.—BOOT FOR ELEVATOR.

Fig. 153 shows a view of a dust-proof elevator made
by Edgar Allen and Co., Ltd., Sheffield, and Figs. 154
and 155 show the boot and hood on a larger scale.

Conveyors.—These machines are used for the continuous
transport of material, and are of various types, such as
worm, scraper, belt, apron, vibrating, and bucket, to
meet the need of transport of various classes of
material.

Worm Conveyor.—The worm or screw conveyor consists of a shaft carrying an endless screw formed by bolting on metal flights. It rotates in a trough having a loose lid, which will lift in case the material as it is carried along should accumulate at any one spot and cause choking, which would damage the screw. During the progress of the material a certain amount of mixing takes place,

FIG. 155.—HOOD FOR ELEVATOR.

depending upon the type of worm employed, the selection of which will be regulated by the nature of the material conveyed. In some cases the worm is made of separate blades or paddles bolted to the shaft, or a continuous helix is made up of cast-iron sections threaded on. The more modern form consists of a spiral attached to the shaft at a few points, as shown in Fig. 156. This is the type of worm employed in the spiral conveyor shown in Fig. 157, made by the above-mentioned firm in all sizes from 4 inches to 24 inches in diameter, and for conveying from 1 ton to 200 tons per hour. These machines are suitable for conveying cements, lime, chalk, coal, ores, flour, phosphates, wheat, barley, seeds, sugar, and other ground material, over distances up to about 100 feet. The nature of the material transported determines the pitch, which on an average is about half the diameter of the screw, and the speed, which is highest for the small sizes, which are driven at about 100 revolutions per minute. The output of a spiral conveyor depends upon the size of the machine, and the power

FIG. 156.—SPIRAL FOR WORM CONVEYOR.

required to drive is obviously a function of the length of the machine, amount of the output in unit time, the efficiency of the worm, and the coefficient of friction of the materials used.

There are various other types of screw conveyors having the worm fixed to the inner surface of the cylinder, which itself revolves, but these are often combined with sifters, mixers, and suchlike machines, which have already been dealt with.

Scraper Conveyor.—In this type of machine, also known as a drag or flight conveyor, the material is pushed along by means of scrapers fixed to an endless rope or chain. In this simple form it finds many applications in factories where sludges have to be transported which would

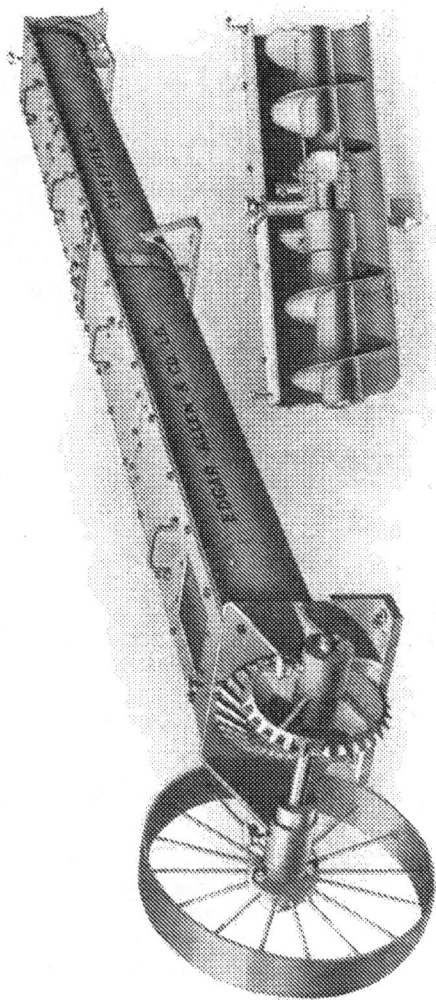

FIG. 157.—SPIRAL CONVEYOR.

rapidly settle and choke a trough through which they flow.

A typical scraper conveyor is the suspended draw type, which consists of a trough along which the material is dragged by flights attached to cross-bars fitted with shoes at the ends for sliding on iron tracks at each side. By this means lateral motion is prevented and the scrapers given the necessary clearance from the sides and bottom of the trough. Fig. 158 illustrates this type as made by Pott, Cassels and Williamson, Motherwell.

FIG. 158.—SCRAPER CONVEYOR.

In another type of machine the wearing shoes are replaced by rollers, which also serve to give the necessary clearances and to reduce friction.

A form of scraper known as a mechanical raker is used for removing salt crystals as they are formed during evaporation in large pans. It consists of a framework suitably braced, and supported on both sides by sliding shoes on tracks provided for that purpose. On this framework, feathering blades are placed at intervals of about 8 feet, and the whole is given a to-and-fro move-

15

ment of about 9 feet, so that there is about 1 foot of overlap in the travel of the blades. The engine used has a cylinder of about 8 inches diameter, and with a 9-foot stroke, which it makes about every two minutes, thus bringing up a load every four or five minutes.

Belt Conveyor.—The belt or band conveyor forms a very efficient means of transportation, for it can be erected in almost any position, and there is a great saving in power over that required for worm conveyors for large quantities over long distances.

It consists of endless belts of rubber (rubber-coated cotton duck), cotton, or metal (wire mesh, etc.), which may be flat or troughed, supported at intervals on rollers, and motion is imparted by a head pulley and the slack taken up by a foot pulley.

Although these machines were originally used for light materials, they are now adapted for heavy work, and for this purpose the rubber belt is designed for rough usage, whereas the cotton belt is more often used for carrying boxes and packages. The capacity of a belt conveyor is determined by its width, which is usually from 10 to 20 inches, and its speed, a troughed belt being capable of transporting two or three times that of a flat belt. Fig. 159 shows the rollers for a three-pulley belt carrier made by Edgar Allen and Co., the upper rollers on the loaded side making the belt into a trough, and the lower roller supporting the belt on the unloaded side. In the ordinary way the material is discharged over the head pulley by centrifugal force, but very often it is required to tap off the material at some intermediate spot. This function is performed by a travelling throw-off carriage such as is shown in Fig. 160, which makes an S-bend in the belt, so that the material is delivered over the top pulley into a hopper placed alongside. With such an arrangement it is necessary to keep the proper tension on the belt, and this is usually effected in the ordinary way of running the belt round a weighted pulley on the

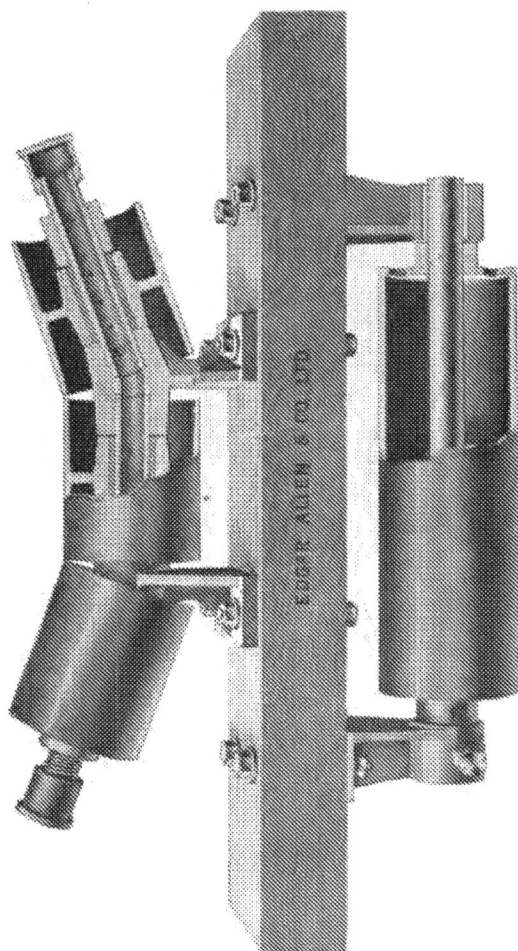

FIG. 159.—THREE-PULLEY BELT CARRIER

FIG. 160.—THROW-OFF CARRIAGE.

unloaded side. Material is fed to the belt by means
of a shoot, which is so adjusted that there is the minimum
relative motion between the material and the belt at the
point of feed, thus reducing the wear on the belt.

Apron Conveyor.—This machine, which is largely used
for handling light packages, consists of light slips of
wood or metal attached to link chains. An ordinary type
of conveyor will not work satisfactorily at a greater
inclination than about 25°; hence there are many
modifications of the belt to enable greater elevations

FIG. 161.—ELEVATOR AND CONVEYOR.

to be used. For packages and such-like the belt may be
provided with raised crossbars which prevent any back-
slip or tendency to overturn.

Fig. 161 shows a conveyor and elevator made by
R. White and Sons, Widnes.

Fig. 162 shows a slat-conveyor made by Pott, Cassel,
and Williamson.

A modification of the belt conveyor is becoming familiar
to most people in the form of the moving staircase,
which is installed in railway stations and other places.

Bucket Conveyor.—Sometimes it is necessary to transport material which from its rough or heated nature is

FIG. 162.—SLAT CONVEYOR.

unsuitable for the belt conveyor. For such purposes the bucket conveyor is used, and it consists of buckets carried on rollers and joined together by a roller chain.

Shaking Conveyors.—These machines have already been dealt with under the heading of shaking sifters.

There are various modifications of these machines, but the underlying principles are the same. They require considerably more power than belt conveyors, but this is somewhat compensated by balancing one shaking conveyor against another, thus securing a comparatively smooth action.

Fig. 163 shows the grasshopper conveyor made by this same firm.

Conveying Liquids.—Liquids are transported from one place to another through pipes under the action of gravity—that is, by maintaining a certain head of liquid. Apart from the means used to obtain a head of the liquid, the main problem in a chemical works consists of selecting a pipe which will remain more or less unaffected by the liquids which flow through it.

For conveying water, pipes are used made of wrought iron, plain or galvanized, cast iron, lead, copper, tin, alloys, and ebonite. Earthenware and cement pipes are commonly used for waste liquids of all kinds. For dilute acids, organic acids, beer, and vinegar, wooden pipes made of staves held together by metal bands are widely used. Lead pipes are useful for resisting corrosion, but they are not satisfactory under heat or pressure, so that lead-lined iron pipes are frequently employed.

Tin pipes are used for conveying liquids which are used in food or for drinking purposes, but a cheaper article is produced by tinning a copper or iron pipe. Copper and brass pipes have as wide an application as any material used in the construction of pipes.

The conveying of strong acids and other chemically active liquids is a problem to the solution of which there have been many attempts. Silicon compounds, in the form of iron alloys or ceramic materials, form the bulk of the acid-resisting substances which are placed on the market under various trade names, such as Ironac,

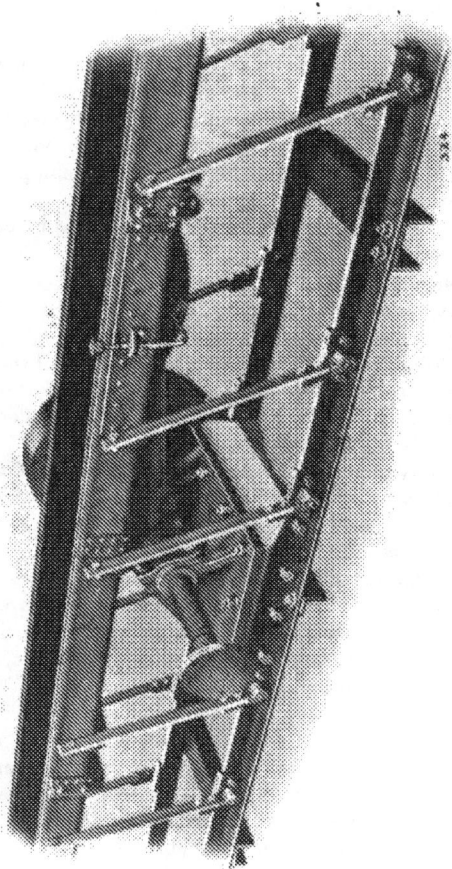

FIG. 163.—GRASSHOPPER CONVEYOR.

Tantiron, Vitreosil, Vitreon, Ceratherm, and Vitresoate stoneware, and intended to replace the regulus metal which has been so commonly used.

Tantiron is the name given to a ferro-silicon alloy manufactured by the Lennox Foundry Company, Ltd., London. It is a hard, close-grained, silvery white alloy, melting at about 1,200° C., which does not rust or oxidize, nor is it attacked by ordinary corrosives to any extent. It can be treated exactly like cast iron, and castings varying from a few ounces to many tons in weight can be made with equal ease. It differs from other non-corrosive alloys in that its resistance to corrosion is general and not specific. Muntz metal, for instance, is not attacked by sea water, and nickel alloys do not rust, but all such metals are easily attacked by acids. Again, truly non-corrosive bodies such as carbides are quite unfitted for the manufacture of plant, as they cannot be cast in the foundry nor be machined. Although in the earlier stages Tantiron was found difficult to machine, and all finished surfaces had to be ground from the rough casting, it can now be drilled, turned, planed, or screwed, and still retains its non-corrosive properties. By immersing weighed samples of Tantiron in different corrosive liquids for periods of one to three days, and carefully weighing the washed and brushed samples at intervals, the table on p. 234 of corrosive actions, giving percentage loss, has been obtained.

As regards physical properties it has been found, as a result of many experiments carefully conducted, to possess practically twice the thermal conductivity of lead and four to five that of stoneware or quartz—an immense gain in either heating or cooling fluids. Its hardness is some fifteen times that of regulus metal, which, together with its lower density, allows lighter and more practical apparatus to be designed than is possible in the case of lead-antimony alloys.

Tantiron has been used in the manufacture of nitric acid and sulphuric acid plants, acid pumps, cocks, valves, pipes, fittings, and various vessels required to withstand exposure to corrosive materials.

	First 24 Hours.	Second 24 Hours.	Third 24 Hours.
Sulphuric acid 98 per cent. ..	0·10	0·02	0·02
,, ,, 30 ,, ..	0·07	nil	nil
Nitric acid 1·4 sp. gr.	0·03	0·01	nil
,, ,, 1·1 ,, ...	0·01	nil	nil
Acetic acid 60 per cent. ..	0·03	0·01	nil
Chromic acid 10 ,, ...	0·07	nil	nil
Tartaric acid 25 ,, ...	0·05	0·03	0·03
Iodine, saturated solution ..	nil	nil	nil
Bromine water, saturated ..	0·01	0·01	nil
Bleaching powder, saturated solution	0·04	0·01	0·01
Copper sulphate, acid	nil	nil	nil
,, ,, alkaline ..	nil	nil	nil
Ferric sulphate solution ..	0·06	nil	nil
Zinc chloride 30 per cent. ..	0·03	nil	nil
Ammonium chloride solution ..	0·05	0·02	0·01
Fused sulphur	0·06	0·01	nil
Fused ammonium nitrate ..	nil	nil	nil

The following physical constants, contrasted with those of cast iron, will be of use to the designer of plant where a non-corrosive metal is required.

	Cast Iron.	Tantiron.
Density	7·3	6·8
Tensile strength, tons per square inch	9 to 10	6 to 7
Transverse strength, 12 inch × 1 inch bars	2,500 pounds	1,600 pounds
Crushing, 1-inch cubes ..	40 tons	34 tons
Melting-point	1,150° C.	1,200° C.
Hardness	1	1·6
Thermal conductivity ..	10	8
Electrical resistance	8	10
Corrosion resistance	1	1,000
Contraction allowance in casting	⅛ inch per foot	1⁄16 inch per foot

Ironac is a similar product manufactured by Houghton's Patent Metallic Packing Co., Ltd., London, which is chiefly used in the construction of special types of nitric acid and sulphuric acid plants. It resists the action of nitric acid and sulphuric acid of all densities, and has sufficient strength, both tensile and transverse, to withstand the necessary handling to which such plants are subjected, and it will resist varying changes of temperature. The conductivity of this material is nearly twenty times that of pottery and similar material, and consequently tubes may be made quite thin in section and cooling effected very rapidly where required.

Owing to the increased efficiency thus gained, a very great saving in space is effected as compared with that required for the old-type pottery installations.

Vitreon ware is made by Shanks and Co., Barrhead, Scotland. It is a pure white, dense body, vitreous throughout, homogeneous in texture, and free from laminations and from iron. It can be used unglazed, as it is vitreous throughout, and has therefore a very low absorption. It resists the action of heat and chemicals equally as well as Berlin porcelain, but it has a very much greater strength. Its compression strength is 24 tons to the square inch, and its tensile strength, calculated by the Nielsen and Garrow formula, is more than 1,800 pounds per square inch, as against 842 pounds for the best German stoneware. A 2-inch diameter pipe with ½-inch walls has successfully withstood a test pressure of 900 pounds per square inch of internal pressure. Pipes of all sizes are made, those of 6 inches diameter being 6 feet long and small bores up to 9 feet in length. Owing to the hardness and fine texture of the material, a fine surface can be obtained by grinding, so that it can be used for the manufacture of acid taps of all descriptions.

Vitreosil is a pure fused silica made by the Thermal Syndicate, Ltd., Wallsend-on-Tyne, and used for the construction of all kinds of appliances used in the acid industries.

It has great resisting powers to high temperatures and the action of chemicals, and is readily made into all kinds of pipes, basins, stills, etc.

Ceratherm is an earthenware composition made by Guthrie and Co., Accrington, and used for the construction of all kinds of chemical machinery. It is unaffected in the slightest degree by corrosive liquids, and, unlike porcelain and earthenware, sudden changes of temperature will not crack it. This material has a much higher thermal conductivity than porcelain or stoneware, and even violent variations of temperature do not crack it, also it can be made of considerable strength without showing the brittle nature of porcelain. Its specific gravity is about one-third that of cast iron, and it has an exceedingly high thermal conductivity and emissivity, and by using a special cement it can be used for lining iron vessels where great strength is required.

Vitreosate is another earthenware composition made by the same firm, and largely used for lining iron pipes and acid cocks.

Elevating Liquids.—Liquids are elevated by the direct action of a plunger pump, centrifugal pump, pulsometer, or hydraulic ram, constructed of non-corrosive material, or indirectly by the use of compressed air in the acid egg system and the Pohlé air lift system.

The Acid Egg.—This apparatus is almost universally used for lifting strongly corrosive liquids, despite the fact of its low efficiency and the labours of chemical engineers to perfect automatic elevators. Its simplicity is a great point in its favour, but its limitation up to the present has been its corrodibility if made of iron or steel, and its weakness if made of earthenware or similar non-corrodible substances.

Fig. 164 is an illustration of an acid egg, as made by The Lennox Foundry Co., in Tantiron, and is formed of two cups joined by their top flanges to form a horizontal cylinder with hemispherical ends. These eggs are filled with the liquid through a pipe which contains a check valve to prevent its return, and air is pumped in through a second pipe having an automatic or manually operated valve. The liquid is then forced out through a third pipe which goes to the bottom of the egg, and at each discharge the liquid is followed by a rush of the air used for raising the liquid and so becomes wasted. Various devices for

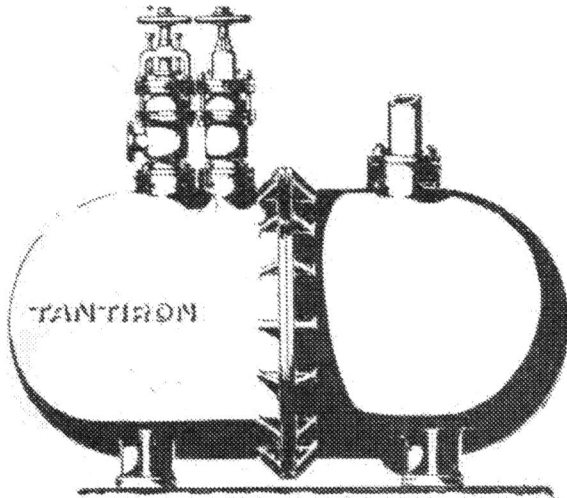

FIG. 164.—TANTIRON ACID EGG.

automatically charging and discharging the eggs have been made and successfully put into operation at acid plants throughout the world.

The Air Lift or Pohlé System.—In this system air under pressure is forced down a pipe within the tubing of a well containing the liquid. The air is broken up into small bubbles, which rise up the tube, accompanied by a certain amount of the liquid, which is discharged at the top in a steady stream. In this system the air escapes, but the amount required is only about half that consumed in the working of an acid egg. The successful working of these pumps demands considerable practical experience, but in general the deeper the submersion of the air pipe, the higher the air pressure, and consequent greater efficiency, which also increases with increase of temperature of the liquid pumped. The absence of moving parts and the freedom from wear makes this pump compare very favourably with positive-acting steam pumps, which suffer from corrosion, except in the case of low lifts of 75 feet and under, when the centrifugal pump is probably more economical.

For practical working it is found that the velocity of the air should not exceed 20 feet per second, that the submersion should be about 1·5 times the lift, measured from the working water level, and the cross-sections of the air tube and the rising tube should be in the ratio of 1 to 6·25. As a rough estimate it takes 1 cubic foot of air to raise 1 gallon of water, but this amount of air can be considerably decreased in the case of an efficient pump. The average air pressure used is 60 pounds per square inch, and at the commencement of operations, owing to the unbroken column of liquid in the tube, a larger pressure is required than is subsequently needed for steady working.

Plunger Pumps.—These pumps exist in all forms and sizes, and are too well known to need any description here. In the design of these pumps for chemical work care should be taken to render all valves easily accessible and to proportion each part to withstand rough usage.

Fig. 165 shows a standard horizontal pump made of Tantiron by the Lennox Foundry Co. specially for acid work. It is a single-acting ram pump with a ram 2 inches in diameter and 6 inches stroke, and when run at 80 revolutions per minute has a capacity of about 280 gallons per hour. The diameters of the suction and delivery pipes are each 1½ inches.

Fig. 166 shows a Tantiron standard vertical pump made for the same duties as the horizontal pump, the dimensions of the cylinder and pipes being exactly the same.

This type of pump is also largely used for pumping wort, molasses, sugar juice, oil, glue, varnish, and other thick liquids. In the case of liquids of heavy density or

FIG. 165.—TANTIRON HORIZONTAL PUMP.

of a sticky nature the output of such pumps is somewhat reduced, and for such cases it is advisable to have large pipe connections, the suction and delivery pipes being of the same diameter as the pump barrel.

Centrifugal Pumps.—The moving of highly corrosive liquids is principally effected by compressed air, but this method is being seriously threatened by the latest types of centrifugal pumps. The manufacture of centrifugal pumps for this purpose, however, raises a variety of very complicated questions, and that is the reason why the development in this direction has been so slow. Lead and regulus metal have been used for strong sulphuric acid, but are not satisfactory for weak acid, and quite useless for nitric and hydrochloric acids or sulphuric acid containing certain common impurities. In addition, all

FIG. 166.—TANTIRON VERTICAL PUMP.

metal pumps are unsuitable for many chemicals and for a large range of solutions of metallic salts. For example, an iron pump could not be used for a solution of chloride of copper. Pumps made from ferro-silicon alloys are not satisfactory when organic acids are used, and the brittleness of the alloy and the difficulty in making suitable castings have prevented its development as a material of which to make pumps.

It is out of the question to use enamelled iron for making pumps, however good the enamel may be, as the enamel, in all probability, will be scratched off the impeller or the casing. In fact, for a number of corrosive liquids no metallic substance is a suitable material from which to make the centrifugal pump, as at every revolution of the impeller the casing and impeller are washed by the

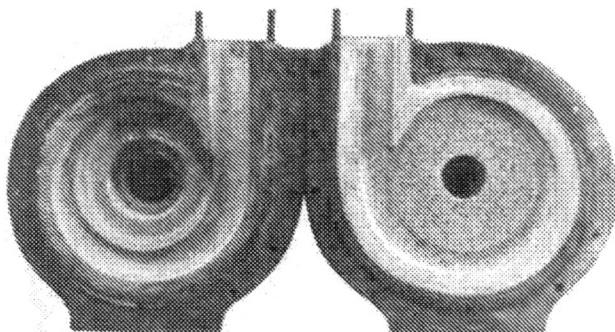

FIG. 167.—CERATHERM BODY IN IRON CASTING.

contained liquid. Porcelain is inert, but is not suitable for the preparation of pumps because of its brittle nature. Stoneware pumps have been produced, but they have the disadvantage that if used with boiling liquids they are liable to crack. They are also very fragile, and are hardly safe to use above 30 pounds pressure or 40 feet acid head. ·

Ceratherm is the material used by Guthrie and Co., Accrington, as the basis for the construction of centrifugal pumps of all kinds.

Taking first of all the case where only low lifts are required and small pumps, a thin casing of Ceratherm

16

FIG. 168.—CERATHERM IMPELLER.

FIG. 169.—CERATHERM PUMP: INTERIOR

is set into an outer casing of cast iron by means of an acid-resisting cement, which will withstand continued treatment with nitric acid, and affixes the lining to the casing in a permanent fashion, as shown in Fig. 167.

Fig. 168 shows an impeller screwed on to a steel shaft which is protected by a Ceratherm boss, after which the Ceratherm is machined and ground accurately. Fig. 169 shows the interior of a pump where the liquids passing through only meet the Ceratherm and do not touch

FIG. 170.—CERATHERM PUMP: SMALL SIZE.

any metal, and in Fig. 170 the complete pump is shown. These pumps are suitable for a number of purposes, and will handle 20 to 100 gallons per minute to a head of 15 to 20 feet without trouble, and they are not affected by boiling corrosive liquids. Stronger pumps are made by using a Ceratherm lining up to $2\frac{1}{2}$ inches in thickness.

No matter how generous the stuffing box of an acid centrifugal pump may be, it is most desirable that the acid at this point should not be under pressure but rather under suction, so as to avoid leaks with high heads,

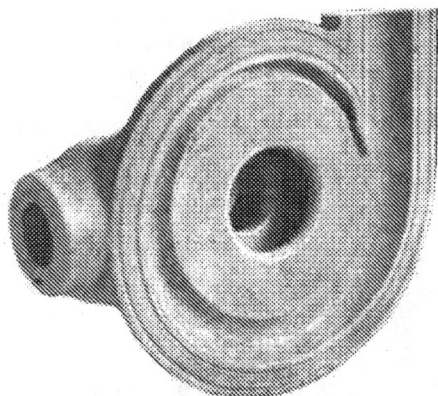

FIG. 171.—CERATHERM PUMP: SUCTION SIDE INTERIOR.

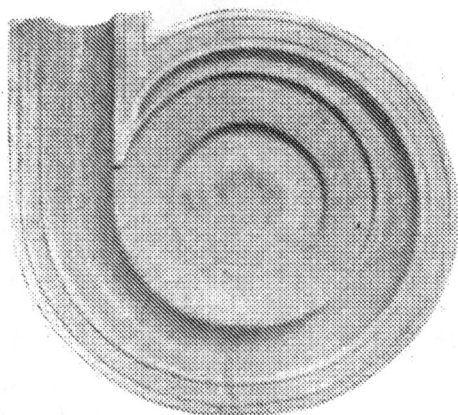

FIG. 172.—CERATHERM PUMP: PRESSURE SIDE INTERIOR.

especially in the case of liquids of high specific gravity.
Fig. 171 shows the interior of the suction side of the
pump containing the feed chamber, inlet, and race, and
Fig. 172 shows the pressure side of the pump containing
the race.

The Ceratherm impellers are tested by running to about
3,000 revolutions, and as the pumps are designed in such
a way as to give the desired head at about 1,000 revolutions, it will be seen that the safety factors in these pumps
are high. The factors are based on a calculation of the
tensile strength, and are probably somewhere in the
neighbourhood of 14 to 15. The distribution of thrust
in a pump of this description is quite different from that
of the normal centrifugal pump, and this is catered for
by the strong casing. It is clear from the above that
the thicknesses of the material found to be desirable
lead to certain modifications in the feed and in the design
of the impellers, but by careful experimentation thoroughly
efficient combinations have been attained. The mechanical efficiency of these pumps is much higher than that
of their only possible competitor in most cases—namely,
the compressed air system. It is not so high as that
attained in metallic pumps, for certain definite reasons—
for instance, 75 per cent. efficiency in a metallic pump
is quite easy to attain, given sufficient quantity. In the
armoured Ceratherm pump a small amount of horse-
power is deliberately spent in preventing the pump from
leaking at the gland, in doing away with troublesome
stuffing boxes, and permitting such contrivances as will
avoid any rubbing surfaces or bearings within the
chemicals. It is fatal practice in chemical pumps to
have bearings running in acids, as this soon causes the
pump to be scrapped.

According to De Laval, until recently 40 per cent. was
considered the highest efficiency for a metal centrifugal
pump of ordinary volute type, and for the quantities
which are generally dealt with in the chemical practice—

namely, below 250 gallons per minute—55 per cent. is as high efficiency as can be attained. It is only when quantities reach the neighbourhood of 500 to 600 gallons per minute that efficiency can be attained up to 70 to 75 per cent. For practical purposes these pumps are at least three times as efficient as the compressed air system, and are greatly superior to metallic pumps which lose their efficiency owing to rapid corrosion. It will be noticed that only single-stage impellers are used, as it has been found that for corrosive liquids, one-stage pumps with very large impellers, which at low revolutions will give the head required in chemical practice, will give the best results.

These pumps are used in bleaching and dyeing works where chemicals must be circulated without any impurity being yielded to the liquors; in acid factories; for the circulation of weak acid in nitric towers; for improved methods of absorbing hydrochloric acid; for electro-chemical purposes, in the wool carbonizing trade, the dope trade, and the fermentation industries.

To maintain freedom from corrosion in the plant actuated by these pumps, armoured Vitreosate piping, another form of earthenware, is used, and armoured Vitreosate acid cocks, of which latter a sample is shown in Fig. 173.

Conveying Gases.—The gases which are met with in the chemical industry may be classified as—(1) Those which are circulated because they are valuable, and (2) those which are exhausted into the air because at present they are of no value. Transportation is effected through pipes which are of various materials, such as iron, in the cast, wrought, galvanized, or sheet form, copper, and for large pipes or flues, bricks, concrete, and similar materials are used.

A chimney is the simplest means of exhausting waste gases to the atmosphere and at the same time providing a draught for promoting the combustion of fuel. but it may be confidently expected that the future will see the abolition of chimneys and the economical use of present waste gases—a subject up to the present sadly neglected. For the removal of noxious gases the size of the chimney is more important than height, which latter is of importance for draught, a height of 400 feet giving a draught pressure equal to about $2\frac{1}{2}$ inches of water.

For low pressures and rarefactions, fans and blowers are used, and for high pressures and rarefactions compressors and vacuum pumps are employed.

' Fig. 173.—Vitreosate Three-Way Tap.

Fans.—These consist of a number of blades fixed on a rapidly revolving shaft, the design of the impeller and casing depending upon the nature of the work to be done. For large volumes of air up to 4 inches water pressure shallow blades are required; for pressures up to 10 inches side plates are provided and the blades are deeper; up to 15 inches the wheel is narrower and deeper and the scroll casing has to be carefully designed; high-pressure fans for cupolas, etc., are provided with a greater number of blades.

For hot or corrosive gases a water-cooled steel fan

may be employed, or the fan may be made of regulus metal or stoneware.

For supplying the air to a producer-gas apparatus the radial-flow fan is frequently employed. It consists of a shaft carrying a spider of six T-iron arms, each having a sheet-iron paddle with annular discs riveted at the sides. The paddle has a radial direction at the inlet, but is curved back at the tip at an angle of about 50 degrees. The side clearance of the impeller is about $1\frac{3}{8}$ inch, and the tip clearance varies from 3 inches at the beak to 12 inches at the discharge orifice. A vortex chamber is provided for by the eccentric setting of the impeller in the casing.

For large and varying volumes of gases these fans are particularly useful, and in spite of losses due to eddy currents and other causes they have an efficiency of from 50 to 70 per cent.

For circulating large volumes of gases at comparatively high pressures a mixed-flow fan, where the inlet flow is axial and the outlet flow radial, such as in the Rateau fan, has come into common use. The shaft of the impeller has a conical boss on which are riveted up to twenty-four specially designed blades having an angle at the tip of about 45 degrees opposite to the direction of rotation, and the vortex chamber is of rectangular cross-section. Delivery is taken from the now usual conical-shaped funnel, and the bearings are of the high-speed type, while carbon rings are provided to prevent leakage at the impeller. This type of fan has a much higher efficiency than the radial-flow type, and with a single impeller can maintain a head of from 35 to 40 inches of water, so that by using several impellers, through which the gas passes in series, large volumes may be passed against high heads.

When a centrifugal fan is run at constant speed the power increases as the pressure falls and the volume of air increases.

Rotary Blower.—This machine is designed to maintain constant blasts of small capacity, and is chiefly used for forge-smiths' fires, furnaces, heating and drying stores, glass blowing, brass smelting and gas exhausting, besides the circulation of gases where high pressures up to 15 pounds per square inch are required. The action is positive, and when run at constant speed the power increases directly with the pressure, while the volume remains practically constant. These machines will run for years without repair, as their mechanism is simple, being designed to scoop the gas at the inlet and discharge it at the outlet, and they are made in large sizes.

Compressors.—These machines are used for pressures considerably above that of the atmosphere, and as such are the development of the last fifty years. All compressors operate by first allowing the gas to flow into a cylinder under its own pressure on the so-called suction stroke of the piston, and then on the return stroke, by closing the inlet valve compressing the gas until the desired pressure is reached, when the delivery valve opens. The following conditions should be fulfilled in a good compressor: (1) On the suction stroke the cylinder should be filled with the maximum mass of gas. (2) On the compression stroke there should be no loss of gas by too late closing of the inlet valves, nor should there be any leakage back, but the whole contents of the cylinder, less the minimum clearance, should be discharged through the outlet valves. (3) The outlet valves should have ample area of opening, should open automatically on the pressure in the receiver being reached in the compression cylinder, and the gas should be discharged at a pressure as little above that in the receiver as possible, as excess pressure causes a rise in temperature, with increase in volume, requiring a corresponding increase in the work necessary to compress and discharge the air. (4) The discharge valves should have sufficient width of seating to ensure their keeping quite tight, so that no loss by leakage back into the cylinder may take place. (5) The valves should be self-adjusting under all speeds and pressures. (6) Highest volumetric efficiency should be obtained. (7) All valves and pistons should be easily accessible for examination and renewal. (8) Wear and tear should be reduced to a minimum.

Compressors are usually of the horizontal type for slow speeds and of the vertical type for high speeds, both requiring very heavy foundations. The horizontal type is more accessible than the vertical, but needs a larger flywheel; but at slower speeds the more even turning of the horizontal type is most marked.

When compressing in a single cylinder to the final pressure desired, the water jacket of a single-stage compressor is not sufficient to extract the heat of compression, and the compound compressor, with intercooler, should always be adopted when the most economical results are required. The rise in temperature not only represents the loss of work in compression, but, should it exceed the ignition temperature of the lubricating oil, an explosion may result. To prevent this, and to obviate the serious loss of efficiency in compression at high terminal pressures which would be caused by the heating of the air, the compression is carried out in stages in a compound compressor, with intercoolers to cool the air between each stage of the compression. The intercooler should be of sufficient capacity to reduce considerably the temperature of the gas before it is admitted to the second-stage cylinder. Provided the work is equally distributed between the cylinders and the intercoolers are properly designed, the final temperature in each cylinder will be the same, and the final temperature of compression very much lower than if the compression were done in one cylinder, with corresponding direct saving in power, as the resistance due to compression is directly proportional to changes in temperature. The compounding of the air cylinders should be so proportioned as to divide the work of

compression equally between them, and thus distribute
the load more equally throughout the stroke, thereby
admitting of an earlier cut-off on the steam cylinders,
with attendant economy in steam consumption.

FIG. 174.—ROBEY COMPRESSOR.

Compressors may be classified according to the manner
in which their inlet and discharge valves are actuated:
(1) Inlet and outlet poppet valves arranged (a) in the

cylinder covers; (b) partly in the cylinder covers and partly in the cylinder walls; (c) with inlet valves in the piston and discharge valves either in the cylinder covers or walls. (2) With large hinged flap valves in the cylinder covers, and actuated similarly to poppet valves. (3) With mechanically operated valves which open and close gradually. (4) With mechanically operated inlet

FIG 175.—VERTICAL OPEN-TYPE AIR COMPRESSOR.

valves and poppet discharge valves. (5) Direct air controlled and balanced inlet and discharge valves. (6) Light automatic disc valves with small lift.

The first four types have a low efficiency and output, owing to the limitations of speed of the pistons due to the heavy valves in motion, and the air-controlled system, although admitting of high speeds and increased efficiency, has not been generally adopted owing to the cost of maintaining the complicated valve gear in good order.

The automatic disc valve is the best valve for practical purposes, as it is silent, simple, and gives great efficiency at high speeds.

Fig. 174 shows a horizontal type two-stage belt-driven air compressor made by Robey and Co., Lincoln, and Fig. 175 a vertical open-type compressor made by the same firm.

FIG. 176.—BELT-DRIVEN THREE-STAGE COMPRESSOR.

Probably the largest field for compressed air is for pressures up to 6 atmospheres, but the rapid progress of the chemical industry has called for higher pressures and for the compression of many other gases. The production of oxygen for welding, etc., of nitrogen for ammonia and nitrates, of argon for electric lamps, depends upon the liquefaction of air demanding up to 200 atmospheres. Hydrogen, chlorine, carbon dioxide, and sulphur dioxide are also stored under pressure.

Fig. 176 shows a belt-driven 3-stage stationary type of small air compressor for simple working made by Peter Brotherhood, Ltd., Peterborough, in which the cylinders are water-cooled and control is effected by an automatic governor.

Fig. 177 shows an oxygen compressor made by this same firm. It is a four-stage machine (2,000 cubic feet per hour, 250 atmospheres at 200 r.p.m.) with forced

FIG. 177.—FOUR-STAGE OXYGEN COMPRESSOR.

lubrication, driven by a two-crank compound steam engine. This type of compressor has lubrication suited to the chemical nature of the gas compressed, and the materials of construction are also adapted to the same end. Generally, the pump is made of bronze of great strength and non-corrodible, except by acetylene and similar gases. There is a minimum number of joints to give trouble, and the piston packing is easily renewable without the use of moulded leather or special piston rings.

Vacuum Pumps.—These machines have been dealt with from time to time in connection with other types of plant already mentioned, and are of the wet and the dry type designed to give what the ordinary engineer terms a perfect vacuum. However, the production of the ionic valve, the latest forms of electric lamps, vacuum vessels for holding liquefied gases, etc., require a high vacuum, which is only limited by the nature of the materials used for construction. For these purposes the high-speed rotary or molecular pump, such as the Gaede, which has a porcelain grooved disc half-immersed in mercury, or the Siemens or Langmuir pumps, which are completely immersed in oil, will give when working alone up to a vacuum equal to $\frac{1}{1000}$ mm. Hg. They are usually worked in series with a large capacity vacuum pump known as a roughing pump and followed by a diffusion pump, which consists of a vessel surrounded by a hot-water bath and containing mercury, which is vaporized continuously, thus providing a wall of vapour through which the attenuated gases diffuse to the rotary pump, a liquid-air cooled trap holding back any straying mercury vapour. By this means much higher vacuums than $\frac{1}{1000}$ mm. Hg. can be obtained on an industrial scale.

CHAPTER IX

APPENDIX

Distillation of Liquid Mixtures.—If the liquids do not mix to any appreciable extent, each exerts its own vapour pressure independently of the other liquids which may be present. The vapour pressure is the sum of the vapour pressures of the liquids contained in the mixture. Such a mixture will boil lower than the lowest boiling constituent, since the sum of the vapour pressures becomes equal to that of the atmosphere at a lower temperature than is required for any one constituent. The vapour of such a mixture will contain all the constituents in the same proportions as the relative vapour pressures of the liquids present. On distillation the distillate will contain all the liquids present in the proportions depending on the relative vapour pressures at the temperature of distillation.

If the liquids are partly miscible, the vapour pressure of the mixture is less than the sum of the vapour pressures of the constituents. The boiling point may be below that of the lowest boiling constituent, or coincident with it, or even above it. The composition of the distillate remains constant so long as there are two layers present, and the effect of distillation is to diminish the lower boiling more rapidly than the higher boiling solution. During this period the boiling point remains constant until one layer has disappeared.

If the liquids are soluble in one another in all proportions, the vapour pressure of the mixture is always less than the sum of the vapour pressures of the constituents at the same temperature. The composition of the vapour from this mixture bears no close relation to the composition of the mixture, but the vapour contains a preponderating amount of the most volatile constituent, and upon this fact rests the possibility of separating such mixtures by fractional distillation. No general relationship exists between the boiling point of such a mixture and the boiling points of the constituents.

Air Compression.—When air is subjected to pressure its volume is reduced and its temperature is raised. If during compression air neither loses heat to, nor gains heat from, any outside source it is said to be adiabatically compressed. In this case the temperature does not remain steady, but rises throughout the operation. In the case where the heat due to compression is removed as quickly as it is formed, so that the temperature of the air remains steady throughout the operation, the compression is said to be isothermal. When the operation is reversed and the air is expanded instead of being compressed, the above holds true, but in the reverse direction —e.g., the volume is increased and the temperature is decreased. The rate of increase of temperature of air during compression decreases as the compression increases, and also—what is very important in practice—it depends upon the initial temperature of the air before compression.

The rate of increase of temperature due to compression increases not only as the initial temperature increases, but also increases throughout the compression when compared with the rates of increase of temperature throughout compression, of air compressed at a lower initial temperature. For a considerable range of temperatures and pressures of air, the relation pressure × volume = constant, provided the temperature is unaltered, holds sufficiently accurately. The effect of increase of temperature is to increase the volume of the air if the pressure is kept unaltered, and hence it follows from the above relation that increase of temperature will increase the pressure of

17

air if the volume is kept unaltered. The final volume or pressure of air under the conditions just mentioned depends directly upon the final absolute temperature of the air—e.g., a certain volume V of gas under pressure P at $t_1°$ C. is heated to $t_2°$ C. If the pressure is kept constant its volume increases from V to $\dfrac{V(273+t_2)}{(273+t_1)}$, and if the volume is kept constant its pressure increases from P to $\dfrac{P(273+t_2)}{(273+t_1)}$. If the Fahrenheit scale is used, then 461 must be used in place of 273. The important facts to be known in connection with air compression are the temperature when any given pressure is reached, and the relative volume of the air at that pressure. It will be seen from the above that as during compression the temperature rises, so the pressure rises also; that is, the back pressure of the air on the piston of the compressor during compression is due in part to the heat generated by the compression. Since air is required to be delivered at a definite temperature and pressure, any rise of the temperature of the air above this definite temperature during compression results in setting up a back pressure which has to be overcome by the prime mover, which is a distinct disadvantage. The required temperature of delivery is usually that of the free air at the intake, hence the ideal compressor should compress free air isothermally. This means that in a compressor the heat produced by compression must be taken away as quickly as it is produced. In any compressor some heat is removed by radiation and by conduction through the metal parts in contact with the air, and usually this operation is assisted by a flow of cold water round the parts wherever possible. The cold-water jacket is more effective on the cylinder head, because that portion is longer exposed to the heated air than any other part; and it should be noted that, apart from other reasons, cooling is a necessity for obtaining proper lubrication and preventing firing in the cylinders. Further, it is obviously an advantage to have a slow-running compressor, for, apart from purely mechanical considerations, the longer the time allowed for the air to cool, the greater will be the cooling effected, and the less the power required for compression. There are two distinct operations in air compression which should be noted—viz., (1) the compression of the air to a given pressure, and (2) the delivery of the air from the cylinder after the given pressure is reached. It will be seen that these operations are the inverse of those occurring in a steam-engine cylinder. In making any calculations as to the h.p. required to compress a given quantity of air to a definite pressure, it is best to take values for adiabatic compression and allow a small percentage for friction of the apparatus, as a part of the friction loss is set off by the reduction of power required due to cooling in actual practice. Assuming there is no clearance in a cylinder of volume v_1 to which air is admitted at a pressure p_1, and that the air is compressed to a volume v_2 at a pressure p_2 adiabatically, the following relations can be determined:

1. Work done by external air in filling the cylinder $= p_1 v_1$ foot-pounds.

2. Work of compression $= 2 \cdot 463 p_1 v_1 \left\{ \left(\dfrac{p_2}{p_1} \right)^{0 \cdot 29} - 1 \right\}$ foot-pounds.

3. Work of expulsion $= p_1 v_1 \left(\dfrac{p_2}{p_1} \right)^{0 \cdot 29}$ foot-pounds.

The total work done is the algebraic sum of these three, and equals $3.463 p_1 v_1 \left\{ \left(\dfrac{p_2}{p_1} \right)^{0 \cdot 29} - 1 \right\}$ foot-pounds, and from this the mean effective pressure (M.E.P.) during the stroke equals $3.463 p_1 \left\{ \left(\dfrac{p_2}{p_1} \right)^{0 \cdot 29} - 1 \right\}$ in pounds per square foot.

The question of clearance cannot be avoided in practice,

but as the compressed gas in the clearance is expanded on the back stroke of the piston, it serves as an additional source of cooling. For pressures above 100 pounds per square inch compressors work in two or more stages, according to the ultimate pressure desired, and intercoolers between the stages are provided to assist in the cooling required. In such cases the stroke in each cylinder except the last stage is a complete compression stroke, without any work of delivery being done, as each cylinder compresses its air into the volume of the cylinder of the next stage. This will modify any calculation for the h.p. required for a multistage compressor. Since in an air compressor the back pressure on the piston is greatest at the end of its stroke, the compressor must be kept in motion by other means than that of the prime mover, and so far this has been effected by heavy flywheels and reciprocating parts. The efficiency of the compressor largely depends upon securing the largest possible mass of air in the first cylinder, which should be at as low a temperature as possible, as it is found that a difference of 5° F. of the air at the intake secures a saving of 1 per cent. The inlet should have an area not less than 50 per cent. of the area of the air piston.

Belt Conveyors.—The power required varies very much with the design, but the following formula may be taken as giving an average result:

$$\text{H.P.} = \frac{(0\cdot15b + 0\cdot07W_1)V + hW_2}{33,000},$$

where V=speed of belt in feet per minute; b=weight of belt; h=height of elevation in feet; W_1=maximum weight of material on the belt at any one time; W_2= weight of material delivered per minute.

Belting.—The power transmitted by any belt:

$$\text{H.P.} = \frac{(T - t)v}{33,000},$$

T = tension in pounds on pulling side; t = tension in pounds on slack sides; v = speed of belt in feet per minute.

Size of pulley and area of contact of belt have no effect on the power transmitted; centrifugal force reduces power by 10 per cent. at 3,000 r.p.m. and 30 per cent. at 5,000 r.p.m.; the arc of contact of belt has considerable influence; the greater the arc, the greater the power transmitted.

A simple rule for belts is that 1 foot per minute of belt speed per inch of width of belt is safely equal to the transmission of 1 watt of electrical energy; add 25 per cent. for light double bands and 60 per cent. for heavy double bands. This fixes a constant for working tensions of 44·24 pounds pull per inch width of belt. The extra pull on the tight side is obtained by dividing the total output in watts by the velocity in feet per minute and multiplying by this number.

Rubber belts transmit 25 to 40 per cent. more power than leather for the same arc of contact, but should not be used for temperatures above 90° F.

Shafting.—For ordinary light shafting carrying pulleys, H.P. = $0·013 \times D^3 \times N$; for shafts carrying gears, H.P. = $0·01 \times D^3 \times N$; for heavy shafting, H.P. = $0·008 \times D^3 \times N$ where D = diameter of shaft in inches and N = speed of shaft in r.p.m.

For wrought-iron shafts, the diameter in inches at the bearings = $5 \times \left(\dfrac{\text{H.P.}}{\text{r.p.m.}} \right)^{\frac{1}{3}}$. Distance between supports, when no power is taken off between, = $5 \sqrt[3]{d^2}$, where d = diameter in inches. In other cases the distance is from 7 to 12 feet.

Maximum safe loads in pounds per square inch on ordinary bearings for shafting: Wrought iron on cast iron = 250; wrought iron on gun-metal or mild steel on cast iron = 300; mild steel on gun-metal = 370; cast steel on gun-metal = 600; flywheel shaft = 250.

Refrigerating Machines.—25 h.p. will cool 15,000 cubic feet of air per hour; 20,000 cubic feet of air saturated at 90° F. contains 42 pounds of water, and at 60° F. contains 17 pounds. The amount of cooling water required for air-compression machines in gallons per minute $= \dfrac{H}{10(T_1 - T_2)}$, where H = heat to be withdrawn from the water, T_1 and T_2 inlet and outlet temperatures respectively.

Low Boiling-Point Liquids.—

				Boiling-Point.	Melting-Point.
Ammonia ..	..	..	..	− 38·5° C.	− 77·34° C.
Carbon dioxide	..	..	..	− 78·2° C.	− 65·0° C.
Ethyl chloride	..	..	..	− 19·5° C.	− 141·6° C.
Liquid air..	..	..	..	− 190° C.	—
Nitrogen ..	..	..	..	− 195·5° C.	− 210·5° C.
Oxygen ..	..	..	..	− 182·7° C.	− 227° C.
Sulphur dioxide	..	..	..	− 10° C.	− 76·1° C.

Freezing Points of Common Salt Brine.—

Specific Gravity at 15° C.	Freezing-Point, ° C.
1·037	− 3·7
1·073	− 7·4
1·111	− 11·0
1·150	− 13·9
1·191	− 17·2

Freezing Points of Calcium Chloride Solutions.—

Specific Gravity at 20° C.	Freezing-Point, ° C.
1·100	− 7·8
1·125	− 10·8
1·150	− 14·2
1·175	− 18·9
1·120	− 24·7
1·225	− 30·8
1·250	− 38·0

Freezing Mixtures.—

Alcohol 77: Snow 73 gives – 30° C.
Alcohol and CO_2 solid gives – 72° C.
Ammonium chloride 30: Water 100 gives – 5·1° C.
Ammonium chloride 25: Snow 100 gives – 15·5° C.
Ammonium nitrate 100: Water 131 gives – 17·5° C.
$CaCl_2$ $2H_2O$ 100: Snow 70 gives – 50° C.
Chloroform and CO_2 solid gives – 77° C.
Ether and CO_2 solid gives – 100° C.
SO_2 liquid and CO_2 solid gives – 82° C.
66 per cent. H_2SO_4 100: Snow 110 gives – 37° C.

Table Showing CaO in Milk of Lime at Varying Density (Mateczel)

Degree Beaumé.	Per Cent. CaO.	100 Litres. Weight Kilos.	100 Litres. CaO Kilos.	Degree Beaumé.	Per Cent. CaO.	100 Litres. Weight Kilos.	100 Litres. CaO Kilos.
10	10·60	125·9	13·3	38	19·72	149·8	29·5
11	11·12	127·4	14·2	39	19·80	149·9	29·6
12	11·65	129·2	15·2	40	19·88	149·9	29·8
13	12·16	130·8	16·1	41	19·95	150·0	29·9
14	12·68	132·6	17·0	42	20·03	150·0	30·1
15	13·20	134·5	18·0	43	20·10	150·0	30·2
16	13·72	136·3	18·9	44	20·16	150·1	30·3
17	14·25	138·2	19·8	45	20·22	150·1	30·4
18	14·77	139·9	20·7	46	20·27	150·1	30·5
19	15·23	141·7	21·6	47	20·32	150·2	30·6
20	15·68	143·6	22·4	48	20·37	150·2	30·7
21	16·10	145·1	23·3	49	20·43	150·3	30·7
22	16·52	146·2	24·0	50	20·48	150·3	30·8
23	16·90	146·9	24·7	51	20·53	150·3	30·9
24	17·23	147·4	25·3	52	20·57	150·4	31·0
25	17·52	147·8	25·8	53	20·62	150·4	31·1
26	17·78	148·1	26·3	54	20·66	150·4	31·1
27	18·04	148·4	26·7	55	20·70	150·5	31·2
28	18·26	148·6	27·0	56	20·74	150·5	31·3
29	18·46	148·8	27·4	57	20·78	150·5	31·3
30	18·67	149·0	27·7	58	20·82	150·5	31·4
31	18·86	149·1	27·9	59	20·85	150·6	31·4
32	19·02	149·2	28·2	60	20·89	150·6	31·5
33	19·17	149·3	28·4	61	20·93	150·6	31·5
34	19·31	149·4	28·7	62	20·97	150·6	31·6
35	19·43	149·5	28·9	63	21·00	150·6	31·6
36	19·53	149·6	29·1	64	21·03	150·7	31·7
37	19·63	149·7	29·3	65	21·05	150·7	31·7

Comparison of Thermometer Scales.

n Degree Celsius $= \frac{4}{5}n$ Degree Réaumur $= 32 + \frac{9}{5}n$ Degree Fahrenheit.
n Degree Réaumur $= \frac{5}{4}n$ Degree Celsius $= 32 + \frac{9}{4}n$ Degree Fahrenheit.
n Degree Fahrenheit $= \frac{5}{9}(n-32)$ Degree Celsius $= \frac{4}{9}(n-32)$ Deg. R.

C.	R.	F.	C.	R.	F.	C.	R.	F.	C.	R.	F.
−20	−16	−4	20	16	68	60	48	140	100	80	212
−19	−15·2	−2·2	21	16·8	69·8	61	48·8	141·8	101	80·8	213·8
−18	−14·4	−0·4	22	17·6	71·6	62	49·6	143·6	102	81·6	215·6
−17	−13·6	1·4	23	18·4	73·4	63	50·4	145·4	103	82·4	217·4
−16	−12·8	3·2	24	19·2	75·2	64	51·2	147·2	104	83·2	219·2
−15	−12	5	25	20	77	65	52	149	105	84	221
−14	−11·2	6·8	26	20·8	78·8	66	52·8	150·8	106	84·8	222·8
−13	−10·4	8·6	27	21·6	80·6	67	53·6	152·6	107	85·6	224·6
−12	−9·6	10·4	28	22·4	82·4	68	54·4	154·4	108	86·4	226·4
−11	−8·8	12·2	29	23·2	84·2	69	55·2	156·2	109	87·2	228·2
−10	−8	14	30	24	86	70	56	158	110	88	230
−9	−7·2	15·8	31	24·8	87·8	71	56·8	159·8	111	88·8	231·8
−8	−6·4	17·6	32	25·6	89·6	72	57·6	161·6	112	89·6	233·6
−7	−5·6	19·4	33	26·4	91·4	73	58·4	163·4	113	90·4	235·4
−6	−4·8	21·2	34	27·2	93·2	74	59·2	165·2	114	91·2	237·2
−5	−4	23	35	28	95	75	60	167	115	92	239
−4	−3·2	24·8	36	28·8	96·8	76	60·8	168·8	116	92·8	240·8
−3	−2·4	26·6	37	29·6	98·6	77	61·6	170·6	117	93·6	242·6
−2	−1·6	28·4	38	30·4	100·4	78	62·4	172·4	118	94·4	244·4
−1	−0·8	30·2	39	31·2	102·2	79	63·2	174·2	119	95·2	246·2
0	0	32	40	32	104	80	64	176	120	96	248
1	0·8	33·8	41	32·8	105·8	81	64·8	177·8	121	96·8	249·8
2	1·6	35·6	42	33·6	107·6	82	65·6	179·6	122	97·6	252·6
3	2·4	37·4	43	34·4	109·4	83	66·4	181·4	123	98·4	253·4
4	3·2	39·2	44	35·2	111·2	84	67·2	183·2	124	99·2	255·2
5	4	41	45	36	113	85	68	185	125	100	257
6	4·8	42·8	46	36·8	114·8	86	68·8	186·8	126	100·8	258·8
7	5·6	44·6	47	37·6	116·6	87	69·6	188·6	127	101·6	260·6
8	6·4	46·4	48	38·4	118·4	88	70·4	190·4	128	102·4	262·4
9	7·2	48·2	49	39·2	120·2	89	71·2	192·2	129	103·2	264·2
10	8	50	50	40	122	90	72	194	130	104	266
11	8·8	51·8	51	40·8	123·8	91	72·8	195·8	131	104·8	267·8
12	9·6	53·6	52	41·6	125·6	92	73·6	197·6	132	105·6	269·6
13	10·4	55·4	53	42·4	127·4	93	74·4	199·4	133	106·4	271·4
14	11·2	57·2	54	43·2	129·2	94	75·2	201·2	134	107·2	273·2
15	12	59	55	44	131	95	76	203	135	108	275
16	12·8	60·8	56	44·8	132·8	96	76·8	204·8	136	108·8	276·8
17	13·6	62·6	57	45·6	134·6	97	77·6	206·6	137	109·6	278·6
18	14·4	64·4	58	46·4	136·4	98	78·4	208·4	138	110·4	280·4
19	15·2	66·2	59	47·2	138·2	99	79·2	210·2	139	111·2	282·2

Temperature, Pressure, and Total Heat of Steam, with Corresponding Vacuum, reduced to a 30-inch Barometer.

Vacuum in Inches.	Absolute Pressure, Pounds per Square Inch.	Temperature Degrees F.	Total Heat H from 0° F.	Vacuum in Inches.	Absolute Pressure, Pounds per Square Inch.	Temperature Degrees F.	Total Heat H from 0° F.
0	14·7	212	1178·6	24·4	2·744	138·2	1156·2
1	14·21	210·4	1178·2	24·6	2·646	136·8	1155·8
2	13·72	208·4	1177·6	24·6	2·548	135·7	1155·4
3	13·23	206·9	1177·1	25·0	2·450	133·8	1154·9
4	12·74	204·9	1176·5	25·2	2·352	132·3	1154·4
5	12·25	203·0	1175·9	25·4	2·254	130·7	1153·9
6	11·76	201·0	1175·3	25·6	2·156	129·1	1153·4
7	11·27	198·7	1174·6	25·8	2·058	127·3	1152·8
8	10·78	196·8	1174·1	26·0	1·960	125·6	1152·3
9	10·29	194·4	1173·4	26·1	1·911	124·6	1152·0
10	9·8	192·5	1172·7	26·2	1·862	123·6	1151·7
10·5	9·555	191·3	1172·4	26·3	1·813	122·7	1151·4
11	9·31	190·0	1172·0	26·4	1·764	121·7	1151·1
11·5	9·065	188·8	1171·6	26·5	1·715	120·7	1150·8
12	8·82	187·1	1171·1	26·6	1·666	119·7	1150·5
12·5	8·575	185·9	1170·8	26·7	1·617	118·6	1150·2
13	8·33	184·7	1170·4	26·8	1·568	117·5	1149·8
13·5	8·085	183·5	1170·0	26·9	1·519	116·4	1149·5
14	7·84	182·0	1169·6	27·0	1·470	115·2	1149·2
14·5	7·595	180·6	1169·1	27·1	1·421	114·0	1148·8
15	7·35	179·1	1168·7	27·2	1·372	112·8	1148·4
15·5	7·105	177·6	1168·2	27·3	1·323	111·6	1148·0
16	6·86	176·0	1167·7	27·4	1·274	110·2	1147·6
16·5	6·615	174·4	1167·2	27·5	1·225	108·9	1147·2
17	6·37	172·8	1166·7	27·6	1·176	107·3	1146·8
17·5	6·125	171·0	1166·2	27·7	1·127	105·9	1146·3
18	5·88	169·3	1165·7	27·8	1·078	104·5	1145·9
18·5	5·635	167·4	1165·1	27·9	1·029	103·1	1145·4
19	5·39	165·6	1164·5	28·0	0·980	101·3	1144·9
19·5	5·145	163·5	1163·9	28·1	0·931	99·6	1144·4
20	4·9	161·5	1163·3	28·2	0·882	97·7	1143·8
20·5	4·655	159·3	1162·6	28·3	0·833	96·1	1143·3
21	4·41	157·1	1162·0	28·4	0·784	94·1	1142·7
21·5	4·165	155·7	1161·6	28·6	0·735	91·8	1142·0
22	3·92	152·2	1160·5	28·6	0·686	89·7	1141·4
22·2	3·822	151·2	1160·2	28·7	0·637	87·4	1140·6
22·4	3·724	150·3	1159·8	28·8	0·588	84·9	1139·9
22·6	3·626	149·1	1159·5	28·9	0·539	82·5	1139·2
22·8	3·528	148·0	1159·2	29·0	0·490	79·3	1138·2
23·0	3·43	146·9	1158·8	29·1	0·441	76·1	1137·2
23·2	3·332	145·7	1158·5	29·2	0·392	72·3	1136·1
23·4	3·234	144·5	1158·1	29·3	0·343	68·8	1135·0
23·6	3·136	143·3	1157·7	29·4	0·294	64·2	1133·6
23·8	3·038	142·1	1157·4	29·5	0·245	59·5	1132·2
24·0	2·94	140·8	1157·0	29·6	0·196	53·3	1130·3
24·2	2·842	139·6	1156·6	29·7	0·147	45·5	1127·9

Specific Gravities of Soda Solutions at 15 C., with Beaume Degree and Percentages of Dry and Crystallized Soda.

Specific Gravity.	Beaume Degree.	Dry Soda.	Cryst. Soda 10 Aq.	1 Cubic Metre Solution contains—	
				Dry Soda (Kilos).	Cryst. Soda (Kilos).
1·007	1	0·67	1·807	6·8	18·2
1·014	2	1·33	3·587	13·5	36·4
1·022	3	2·09	5·637	21·4	57·6
1·029	4	2·76	7·444	28·4	76·6
1·036	5	3·43	9·251	35·5	95·8
1·045	6	4·29	11·570	44·8	120·9
1·052	7	4·94	13·323	52·0	140·2
1·060	8	5·71	15·400	60·5	163·2
1·067	9	6·37	17·180	68·0	183·3
1·075	10	7·12	19·203	76·5	206·4
1·083	11	7·88	21·252	85·3	230·2
1·091	12	8·62	23·248	94·0	253·6
1·100	13	9·43	·25·432	103·7	279·8
1·108	14	10·19	27·482	112·9	304·5
1·116	15	10·95	29·532	122·2	329·6
1·125	16	11·81	31·851	132·9	358·3
1·134	17	12·61	34·009	143·0	385·7
1·142	18	13·16	35·493	150·3	405·3
1·152	19	14·24	38·405	164·1	442·4

Liquids.

Substance.	Specific Gravity (Water = 1·00).	Density in Pounds per Cubic Foot.	Specific Heat (Water = 1·00).	Boiling-Point in Degrees Cent.
Alcohol	0·791	49	0·673	78
Benzine	0·85	53	0·395	—
Ether	0·723	45	0·516	35
Mercury	13·596	848	0·033	357
Turpentine (oil of)..	·0·865	54	0·463	160
Water (almost boiling)	0·958	60	1·013	100

Solids.

Substance.	Specific Gravity (Water = 1·00).	Density in Pounds per Cubic Foot.	Specific Heat (Water = 1·00).	Melting-Point in Degrees Cent.
Aluminium (cast) ..	2·6	162	0·212	625
Antimony	6·7	417	0·051	435
Arsenic	5·9	368	0·081	—
Bismuth	9·82	613	0·031	260
Brass	8·4	525	0·094	900
Cadmium	8·6	539	0·057	320
Charcoal	0·36	22	0·241	—
Coal (anthracite) ..	1·43	89	0·241	—
Cobalt	8·5	530	0·161	1500
Coke	1·0	62	0·203	—
Copper .·. ..	8·8	550	0·092	1090
Fluorspar	3·15	196	—	900
Glass	2·89	181	0·198	1100
Gold	19·3	1200	0·032	1050
Ice (at 0° C.) ..	0·92	57	0·504	0
Iridium	22·4	1400	0·033	1950
Iron (cast)	7·2	451	0·130	1100
Iron (wrought) ..	7·7	485	0·114	1600
Lead	11·4	710	0·031	325
Limestone	3·16	197	0·217	—
Magnesium.. ..	1·74	109	0·250	500
Manganese	8·0	499	0·122	—
Nickel	8·7	542	0·107	1500
Oak	0·86	54	0·57	—
Palladium	11·4	710	0·059	1500
Pine..	0·55	34	0·65	—
Platinum	21·5	1340	0·032	1775
Silver	10·5	653	0·056	950
Steel	7·85	490	0·116	1500
Sulphur	2·07	127	0·203	115
Thallium	11·8	736	0·034	290
Tin	7·3	455	0·056	230
Zinc..	7·12	445	0·095	415

INDEX

A

ABSORPTION system of cold storage, 209
Acid egg, 236
— retort, nitric, 159
Aerial wire ropeways, single wire, 213
— — — double wire, 216
Air circulation system of cold storage, 209
Air compression, 257
— drying, 103
— lift system, 237
— separators, 49
— — gravity leg, 50
— — "Stag," 51
Ammonia compressor, 205
Apron conveyor, 229
Armoured ceratherm pumps, 241
Aspinall evaporating pan, 125
"Atlas" pebble grinding mill, 31

B

Bag filter, 70
Ball mill, 29
Bearing for centrifugal spindle, 86
Belt conveyor, 226
— — power for, 260
Belting, 260
Blower, rotary, 249
Boiler, corrosion, 180
— foaming, 181
Brine-pipe system, 208
Bucket conveyor, 229
— — "Weston" centrifugal, 84
Buhrstone mill, 24
By-product coke oven retort, 159

C

Calciner, rotary, 169
Can ice system, 206
Cane-juice subsider, 57
CaO in milk of lime, 263
Carbon dioxide compressor, 206
Cast-iron calandria, 129
Cell ice system, 208
Central screw closing of filter press, 81
Centrifugal machines, 84

Centrifugal machines, basket for, 84
— — basket linings, 85
— — bearings for spindle, 86
— — electric-driven, 90
— — friction pulley for, 90
— — interlocking gear, 96
— — "Weston" type, 84
— — dressing, 44
Centrifugal pumps, 239
Ceratherm ware, 236
Chamber kiln, 167
Chamber press, 75
Chaser, 10
Climbing film vacuum pan, 135
Coal-gas retorts, 160
Coffey still, 150
Cold storage system:
— — — air circulation, 209
— — — brine pipe, 208
— — — direct expansion, 209
Column still, 145
Combination tube mill, 37
Comparison of thermometer scales, 264
Compression of air, 257
Compressors:
— ammonia, 205
— carbon dioxide, 206
— oxygen, etc., 253
Cone paint mill, 65
Continuous cone vacuum dryer, 115
Continuous still, 147
Continuous water-softening plant, cylindrical, 188
— — — rectangular, 185
Control of temperature, 194
— — of dye vessel, 200
— — of gas producer, 201
— — of jacketed pan, 199
— — of spinning rooms, 202
— — of still, 198
Control steam valve, 194
Conveying gases:
— — blowers, 249
— — chimneys, 246
— — compressors, 249
— — fans, 247
— liquids:
— — acid egg, 236
— — pipes, 231

Conveying liquids:
— — Pohlé system, 237
— solids:
— — bucket elevators, 217
— — runways, 212
— — tipping waggons, 212
— — wheelbarrows, 211
— — wire ropeways, 213
Conveyors:
— apron, 229
— belt, 226
— bucket, 229
— scraper, 223
— shaking, 231
— worm, 222
Corrosion of boilers, 180
Cracker, 10
Crusher, fine rotary, 10
— jaw-, 1
Crushing rolls, 5
— — high speed, 6
— — fine, 8
Crutcher, 68

D

Direct heat evaporators, 118
Discharge valve, Lassen-Hjort, 188
Disintegrators, 14
— sifter for, 18
— fixing of, 20
Distillation of liquid mixtures, 256
Double mixer for semi-liquids, 66
"Dowson" gas producers:
— — — pressure, 161
— — — suction, 163
— — — bituminous, 163
Drag conveyor, 223
Dresser, powder, 44
Dressing machine, centrifugal, 47
Dryer, mixer, and ball mill, 115
Dryers, non-vacuum:
— — combination, 101
— — "Firman," 100
— — "Hersey," 101
— — rotary, 99
— — Sturtevant, 107
— vacuum:
— — continuous cone, 115
— — drum, 112
— — "Johnstone," 114
— — rotary, 114
— — shelf, 109

E

Edge runner mill, 10
— — — granite, 13
— — — iron, 12
— — — overhead-driven, 14
Electric control, "Isothermal," 197

Electro-magnetic separating machines, 52
Elevating liquids:
— — acid egg, 236
— — Pohlé system, 237
Elevators, 217
Evaporating pan, "Aspinall," 124
— — open type, 121
— — "Wetzel," 125
Evaporators, 118
— direct heat, 119
Extraction plant, 152

F

Fans:
— mixed flow, "Rateau," 248
— radial, 248
Filter, bag, 70
Filter press:
— — frame type, 72
— — — — plates for, 74
— — recessed type, 75
— — — — clips for, 78
— — — — double cloth system, 79
— — — — plates, 79
— — methods of closing, 79
— — — — central screw, 81
— — — — hydraulic, 81
— — — — pneumatic, 81
— — — — rack and pinion, 81
— — methods of feeding, 82
"Firman" dryer, 100
Flight conveyor, 223
Flue heater, 98
Freezing-point of brine, 262
— — CaCl₂, 262
— mixtures, 263
Friction pulley for centrifugals, 90
Furnaces:
— "Harris," 171
— — shafts for, 172
— "H. H." mechanical, 178
— muffle, 169
— regenerative, 170
— reverberatory, 170
— roasting, 171

G

Gas compressors, 254
— conveyance, 246
— retorts, coal, 160
— — "Dowson" bituminous, 163
— — — pressure, 161
— — — suction, 163
— — hydrogen, 166
Gas valves, "Isothermal," 203
Grainer, 121

Grasshopper conveyor, 284
Grizzly, 40

H

Hand-driven portable screen, 42
Harris roasting furnace, 171
— — — shaft, 172
Heater, flue, 98
"Hersey" dryer, 101
"H. H." mechanical roasting furnace, 178
Horizontal mixer, 65
Hydraulic closing of filter press, 81
Hydrogen retorts, Lane process, 166

I

Ice making:
— — can ice, 206
— — cell ice, 208
— — plate ice, 208
Interlocking gear for centrifugals, 96
Intermittent water-softening plant, 184
Ironac, 231
"Isothermal":
— control apparatus, 197
— gas valve, 203
— steam valve, 195
— superheated steam valve, 199
— thermometer, 196

J

Jaw-crusher, 1
Jet condenser, 133
"Johnstone" dryer, 114

K

"Kestner" evaporators:
— — climbing film, 135
— — falling film, 137
— — quadruple effect, 139
— — salting type, 139
Kilns:
— chamber, 167
— lime, 167
— rotary, 168

L

Lane hydrogen retorts, 166
Lassen-Hjort water plant:
— — — cylindrical, 190
— — — rectangular, 185
Levigating mill, 56
— plant, 61
Lightfoot refrigeration system, 204

Lime kiln, 167
Lime-soda water softening plant:
— — — — continuous, 185
— — — — intermittent, 184
Liquid mixtures, distillation, 256
Linings for "Weston" basket, 85
Low boiling-point liquids, 262
Lubricating oil still, 158

M

Machines:
— centrifugal, 84
— — dressing, 44
— crutching, 68
— electro-magnetic separating, 52
— mixing, 61
— vibration, 45
Magnetic pulley, 55
Measuring apparatus, Lassen-Hjort, 185
Mechanical raker, 225
Methods of closing filter press, 81
— feeding filter press, 82
Milk of lime, 263
Mill:
— ball, 29
— Buhrstone, 24
— combination tube, 37
— cone paint, 65
— edge runner, 10
— — — granite, 13
— — — iron, 12
— levigating, 56
— overhead-driven, 14
— pebble, 31
— pug, 64
— putty, 62
— roller, 27
— triple roller, 28
— tube, 34
— vertical runner, 26
Mixer:
— double, 67
— horizontal, 65
— open drum, 67
— undergeared, 68
Mixing machinery, 61
Muffle furnace, 169
Multiple effect vacuum pan, 133
"Multiplex" evaporator, 141

N

"Newaygo" screen, 49
Nitric acid retort, 159

O

Open drum mixer, 67
— evaporating pan, 121

P

Paint mill, 65
Pans:
— "Aspinall," 125
— steam-jacketed, 122
— vacuum, 125
— "Wetzel," 125
Pebble grinding mill, 31
Permutit, 190
Petroleum stills, 154
Plant:
— extraction, 152
— levigating, 61
— lime-soda, continuous, 185
— — intermittent, 184
— — Lassen-Hjort, cylindrical, 190
— — — rectangular, 185
— Permutit, 190
Plate ice system, 208
Pohlé air system, 237
Portable hand-driven screen, 42
Pot still, 159
Powder dresser, 44
Press filter, 71
— — chamber, 75
— — frame, 72
— — — plates, 74
— — recessed type, 77
— — — — clips, 78
— — double cloth system, 78
— — — — plates, 79
— — methods of closing:
— — — — central screw, 81
— — — — hydraulic, 81
— — — — pneumatic, 81
— — — — rack and pinion, 81
— — methods of feeding, 82
Pug mill, 64
Pulley, magnetic, 55
Pumps, acid, centrifugal, 239
— — plunger, 238
— vacuum, Gaede, 255
— — Langmuir, 255
— — mercury, 255
— — Siemens oil, 255
Putty mill, 62

Q

Quadruple effect evaporator, 139

R

Rack and pinion closing of filter press, 81
Rectangular water-softening plant, 185
Rectifying still, 147

Reels, sifting, 43
Refrigerating machines, 203, 262
— — ammonia, 204
— — — absorption system, 209
— — carbon dioxide, 206
Regenerative furnaces, 171
Retorts:
— by-product coke-ovens, 159
— coal-gas, 160
— "Dowson" bituminous, 163
— — pressure, 161
— — suction, 163
— hydrogen, 166
— nitric acid, 159
Reverberatory furnaces, 190
Roasting furnace, 171
Roller mills, 27
— — triple, 28
Rolls, crushing, 5
— fine, 8
— high-speed, 6
Rotary blower, 249
— calciner, 169
— fine crusher, 10
— vacuum dryer, 114
Runway for mine, 212

S

Salting type evaporator, 139
Scraper conveyor, 223
Screen, "Newaygo," 49
— portable hand-driven, 42
Self-recording hygrometer, 104
— — chart, 106
Separation by water, 56
Separators, air, 49
— electro-magnetic, 53
— gravity leg, 51
— "Stag," 5
Settling tank, 57
Shafting, 261
Shafts for roasting furnace, 172
Shaking sifters, 47
Sifters for disintegrators, 18
Sifting reels, 43
Slat conveyor, 229
Soap crutcher, 68
Softening of water, 183
— — plant, continuous, 185
— — — intermittent, 184
— — — lime-soda, 184
— — — Lassen-Hjort, cylindrical, 190
— — — — discharge valve, 188
— — — — measuring apparatus, 185
— — — — rectangular, 185
— — Permutit, 190

Spindle, centrifugal bearing, 86
"Stag" ball mill, 33
Stamps, 38
Steam-jacketed kettle, 123
— pans, 122
— temperature, pressure, and total heat, 265
— valve, "Isothermal," 195
Still:
— Coffey, 150
— column, 145
— continuous, 147
— lubricating oil, 158
— petroleum, 154
— pot, 159
— rectifying, 147
— tar, 158
"Sturtevant," drying system, 103
— hygrometer, 104
— — guide chart, 104
— triple duct dryer, 107
Superheated steam valve, 199
Surface condenser, 133
Synchronous speeds for centrifugals, 86

T

Tables:
— common liquids, 266
— — solids, 267
— comparison of thermometer scales, 264
— specific gravity of soda solutions, 266
— total heat of steam, 265
Tantiron, 233
— pumps, 238
Tar stills, 158
Thermometer, "Isothermal," 196
Three-way tap, 247
Throw-off carriage, 226
Tilting kettle, 123
Tipping waggons, end, 211
— — side, 212
Triple duct dryer, 107
— roller mill, 28
Trommel, 40
Tube mills, 34
Typical guide chart, 106

U

Undergeared mixer, 68
"Universal" cone mill, 66

V

Vacuum dryers:
— — continuous cone, 115
— — drum, 112

Vacuum dryers:
— — "Johnstone," 114
— — mixer and ball mill, 115
— — rotary, 114
— — shelf, 109
— pan:
— — cast-iron calandria, 129
— — copper, 127
— — jet condenser, etc., 130
— — "Kestner" climbing film, 135
— — — falling film, 137
— — — quadruple effect, 139
— — — salting type, 139
— — multiple effect, 133
— — "Multiplex," 141
— — Torricellian condenser, etc., 130
— pumps, diffusion, 255
— — dry, 130
— — Gaede, 255
— — Langmuir, 255
— — Siemens oil, 255
— — wet, 130
Valve, "Isothermal" gas, 203
— — steam, 195
— — superheated steam, 199
Vertical runner mill, 26
Vibration machines, 45
Vitreon ware, 235
Vitreosate, 236
Vitreosil, 235

W

Warm air drying, 103
Water separation, 56
— softening plant:
— — — — discharge valve, 188
— — — — Lassen-Hjort, cylindrical, 188
— — — — lime-soda, continuous, 185
— — — — — intermittent, 184
— — — — measuring apparatus, 185
— — — — Permutit, 190
— — — — rectangular, 185
— — treatment, 179
— — chemistry of process, 183
— — impurities, 181
"Weston" centrifugals, 84
— — basket, 84
— — basket linings, 84
— — electric-driven, 91
— — friction pulley, 90
— — interlocking gear, 96
— — spindle bearing, 85
— — water-driven, 91
"Wetzel" evaporating pan, 125
Worm conveyor, 222

CPSIA information can be obtained
at www.ICGtesting.com
Printed in the USA
BVHW030802250521
608089BV00004B/18

9 781446 069424